T0212646

Modern Birkhäuser Classics

Many of the original research and survey monographs in pure and applied mathematics published by Birkhäuser in recent decades have been groundbreaking and have come to be regarded as foundational to the subject. Through the MBC Series, a select number of these modern classics, entirely uncorrected, are being re-released in paperback (and as eBooks) to ensure that these treasures remain accessible to new generations of students, scholars, and researchers.

Victor P. Pikulin
Stanislav I. Pohozaev

Equations in Mathematical Physics

A practical course

Translated by Andrei Iacob

Reprint of the 2001 Edition

 Birkhäuser

Victor P. Pikulin†
Moscow Power Engineering Institute (MPEI)
Russia

Stanislav I. Pohozaev
Steklov Institute of Mathematics
Russian Academy of Sciences
Gubkina str. 8
119 991 Moscow
Russia

ISBN 978-3-0348-0267-3   e-ISBN 978-3-0348-0268-0
DOI 10.1007/978-3-0348-0268-0
Springer Basel Dordrecht Heidelberg London New York

Library of Congress Control Number: 2011942991

Mathematics Subject Classification (2010): 35Qxx, 35-01, 35Jxx, 35Kxx, 35Lxx

© Springer Basel AG 2001
Reprint of the 1st edition 2001 by Birkhäuser Verlag, Switzerland
Originally published as Physics-Matematicheskaya Literatura by Izdatel'skaya firma, Russian Academy of Sciences.
Translated from the Russian by A. Iacob

This work is subject to copyright. All rights are reserved, whether the whole or part of the material is concerned, specifically the rights of translation, reprinting, re-use of illustrations, recitation, broadcasting, reproduction on microfilms or in other ways, and storage in data banks. For any kind of use, permission of the copyright owner must be obtained.

Printed on acid-free paper

Springer Basel AG is part of Springer Science+Business Media
(www.birkhauser-science.com)

Contents

Preface

This handbook is addressed to students of technology institutes where a course on mathematical physics of relatively reduced volume is offered, as well as to engineers and scientists. The aim of the handbook is to treat (demonstrate) the basic methods for solving the simplest problems of classical mathematical physics.

The most basic among the methods considered here is the superposition method. It allows one, based on particular linearly independent solutions (solution "atoms"), to obtain the solution of a given problem. To that end the "supply" of solution atoms must be complete. This method is a development of the well-known method of particular solutions from the theory of ordinary differential equations. In contrast to the case of ordinary differential equations, where the number of linearly independent solutions is always finite, for a linear partial differential equation a complete "supply" of solution atoms is always infinite. This infinite set of solutions may be discrete (for example, for regular boundary value problems in a bounded domain), or form a continuum (for example, in the case of problems in the whole space). In the first case the superposition method reduces to the construction of a series in the indicated solution atoms with unknown coefficients, while in the second case the series is replaced by an integral with respect to the corresponding parameters (variables).

This first step leads us to the general solution of the associated homogeneous equation under the assumption that the set of solution atoms is "complete." The next step – like in the case of ordinary differential equations–is to find the requisite coefficients from the data of the problem.

Thus, the basic bricks of the superposition method are the solution "atoms." A main tool for finding such solutions is the method of separation of variables (assuming null boundary conditions). Unfortunately, this method is applicable only in the case of domains that possess a certain symmetry (for example, in the case of a disc, rectangle, cylinder, or ball). In the case of domains with a complicated structure finding such solution "atoms" is no longer possible. One can then resort to various approximate methods for solving boundary value problems, which is beyond the scope of the present handbook.

Another method discussed in the book is the method of conformal mappings, which allows one to reduce the solution of, say, the Dirichlet problem for the Laplace equation in a complicated domain, to the consideration of the Dirichlet problem in a simpler domain.

Alongside these methods we consider integral transform methods (Fourier, Laplace, Hankel) for nonstationary problems, which are also based on the linear superposition method.

We also devote place in the handbook to a group of problems connected with equations of 4th order. We consider boundary value problems for the biharmonic equation (in a disc and in a half-space) and for a nonhomogeneous equation of 4th order (in a disc).

To supplement our treatment of methods for solving elliptic, hyperbolic and parabolic problems, each chapter ends with problems for independent study and answers to them. We note that in chapters 2 and 3 we give problems also for hyperbolic and parabolic equations of 4th order, with hints for solution.

Let us emphasize that because this book is addressed to readers that are familiar only with differential and integral calculus and some methods for the integration of ordinary differential equations, we are not discussing questions pertaining to the existence of solutions and their membership in appropriate functional spaces.

The entire exposition is formal in nature; if desired, it can be readily given a rigorous mathematical meaning by enlisting the general theory of partial differential equations. The reader interested in a deeper study of the methods demonstrated here or other methods is referred to the bibliography, which contains a number of well-know collections of problems. In particular, as a handbook we recommend S. Farlow's book [22].

Introduction

Many physical processes in fields of science and technology such as mechanics, heat physics, electricity and magnetism, optics, are described by means of partial differential equations. The majority of the equations of mathematical physics itself are partial differential equations. In contrast to ordinary differential equations, in which the unknown function depends on only one variable, in partial differential equations the unknown function depends on several variables (for example, the electric field vector $\vec{E}(x,y,z,t)$ depends on the space variables x,y,z and the time t).

Among the fundamental equations of mathematical physics one can list

- the three-dimensional Laplace equation

$$\Delta u = 0, \quad \text{or} \quad u_{xx} + u_{yy} + u_{zz} = 0;$$

- the wave equation

$$u_{tt} = a^2 \Delta u;$$

- the heat equation

$$u_t = a^2 \Delta u.$$

In these equations the unknown function $u = u(x,y,z)$, or $u = u(x,y,z,t)$, depends on several variables. The number of independent variables is determined by the dimension of space in which the physical process under study takes place, to which one adds the space variable (in the case of nonstationary processes). A wide class of equations is constituted by the linear partial differential equations for functions that depend on two variables. These equations can be written in the form

$$a_{11}u_{xx} + 2a_{12}u_{xy} + a_{22}u_{yy} + b_1 u_x + b_2 u_y + cu = f, \tag{I.1}$$

where a_{11}, a_{12}, a_{22}, b_1, b_2, c, f are given functions of the independent variables x and y and u is the unknown function.

Any linear second-order partial differential equation of the form (I.1) belongs, under the assumption that $a_{11}^2 + a_{12}^2 + a_{22}^2 \neq 0$ at a point (x,y), to one of the three types: (a) elliptic; (b) hyperbolic; (c) parabolic (at that point (x,y)).

The equations of elliptic type describe steady processes and are defined by the condition $a_{12}^2 - a_{11}a_{22} < 0$.

The equations of hyperbolic type describe wave processes and are defined by the condition $a_{12}^2 - a_{11}a_{22} > 0$.

The equations of parabolic type describe processes of heat propagation, diffusion (and some other), and are defined by the condition $a_{12}^2 - a_{11}a_{22} = 0$.

Let us remark that in equations of hyperbolic and parabolic type one of the variables, denoted t, plays the role of the time variable.

Thus, the equations $u_{xx} + u_{yy} = 0$, $u_{tt} = a^2 u_{xx}$, and $u_t = a^2 u_{xx}$ are of elliptic, hyperbolic, and parabolic type, respectively.

The concept of a solution. A (*classical*) *solution* of a partial differential equation is a function (possessing all the derivatives appearing in the equation) which upon substitution in the given equation makes it into an identity with respect to the independent variables in the domain under consideration. For example, the function $u(x,t) = sinx \sin(at)$ is a solution of the equation $a^2 u_{xx} - u_{tt} = 0$ in the domain $-\infty < x, t < \infty$. Let us notice that an equation has many different solutions. This fact is observed already in the study of ordinary differential equations. In the case of partial differential equations the set of solutions is considerably wider. For example, the set of solutions of the equation $y'' = 0$ is described by the formula $y = C_1 x + C_2$, where C_1 and C_2 are arbitrary constants; by contrast, the set of solutions of the equation $a^2 u_{xx} = u_{tt}$ is given by the formula $u(x,t) = f(x - at) + g(x + at)$, where f and g are arbitrary twice differentiable functions.

Let us emphasize that the concept of a solution introduced above is not the only one possible. A well-developed theory of so-called generalized (with various meanings) solutions exist at the present time. However, in our elementary course we will confine ourselves to the classical definition given above.

Problem posing. In order to single out a unique solution among the set of solutions of a given equation, it is necessary to impose supplementary conditions. These conditions are of various kinds, depending on the type of equation studied. In this way we are led to posing problems for partial differential equations. In other words, a problem consists of an equation and supplementary conditions.

In the case of unsteady processes studied in the whole space it is necessary to give initial conditions. In this way we arrive at the *Cauchy problem*. Typical examples of this problem are the following:

$$\begin{cases} u_{tt} = a^2 u_{xx}, & -\infty < x < \infty, \quad t > 0, \\ u(x,0) = f(x), \quad u_t(x,0) = g(x), & -\infty < x < \infty; \end{cases}$$

$$\begin{cases} u_t = a^2 u_{xx}, & -\infty < x < \infty, \quad t > 0, \\ u(x,0) = f(x), & -\infty < x < \infty; \end{cases}$$

$$\begin{cases} u_{tt} = a^2 \Delta u, & -\infty < x, y, z < \infty, \quad t > 0, \\ u(x,y,z,0) = f(x,y,z), & -\infty < x, y, z < \infty, \\ u_t(x,y,z,0) = g(x,y,z), & -\infty < x, y, z < \infty; \end{cases}$$

$$\begin{cases} u_t = a^2 \Delta u, & -\infty < x, y, z < \infty, \quad t > 0, \\ u(x,y,z,0) = f(x,y,z), & -\infty < x, y, z < \infty. \end{cases}$$

The Cauchy problem already has a unique solution (under natural conditions).

If, however, the physical process is considered in a bounded domain of space, then we are led to boundary value problems for stationary phenomena and mixed problem for nonstationary phenomena. For example, when studying the oscillations of a string clamped at its ends one obtains the mixed problem

$$\begin{cases} u_{tt} = a^2 u_{xx}, & 0 < x < l, \quad 0 < t < \infty, \\ u(0,t) = 0, \quad u(l,t) = 0, & 0 \le t < \infty, \\ u(x,0) = f(x), \quad u_t(x,0) = g(x), & 0 \le x \le l. \end{cases}$$

Other forms of boundary conditions are also possible.

In a similar manner one poses the boundary value problem for the wave equation in the three-dimensional case (Ω is a bounded domain with boundary $\partial\Omega$):

$$\begin{cases} u_{tt} = a^2 \Delta u & \text{in the cylinder } \Omega \times (0,t) \text{ with } t > 0, \\ u = 0 & \text{on the lateral surface } \partial\Omega \times (0,t), \\ u = f(x,y,z), \quad u_t = g(x,y,z) & \text{on the lower base } (t = 0) \text{ of the cylinder.} \end{cases}$$

Further, the process of heat propagation in a rod of length l is described by the one-dimensional heat equation

$$u_t = a^2 u_{xx}, \qquad 0 < x < l, \quad 0 < t < \infty,$$

where $u(x,t)$ is the temperature in the rod at time t in the point x. We shall assume that the endpoints $x = 0$ and $x = l$ are maintained at the temperature zero at all time, and a temperature profile $\varphi(x)$ is given at the initial moment of time. Then we obtain the following mixed problem for the heat equation:

$$\begin{cases} u_t = a^2 u_{xx}, & 0 < x < l, \quad 0 < t < \infty, \\ u(0,t) = 0, \quad u(l,t) = 0, & 0 \le t < \infty, \\ u(x,0) = \varphi(x), & 0 \le x \le l. \end{cases}$$

Instead of specifying the temperature at the endpoints (boundary) of the rod, one can specify the temperature of the ambient medium or the heat flux through the boundary.

The boundary value problem for the heat equation in the whole space is formulated in analogous manner:

$$\begin{cases} u_t = a^2 \Delta u & \text{in the cylinder } \Omega \times (0,t) \text{ for } t > 0, \\ u = 0 & \text{on the lateral surface } \partial\Omega \times (0,t), \\ u = f(x,y,z) & \text{on the lower base } (t = 0) \text{ of the cylinder.} \end{cases}$$

In addition to physical phenomena that evolve in space and time, there exist phenomena which do not change with time. Most of these phenomena are described by elliptic boundary value problems. In contrast to the hyperbolic wave equation and the parabolic heat equation, elliptic boundary value problems require no initial conditions. For them only boundary conditions need to be imposed. The following three types of boundary conditions are the most important ones:

(1) boundary condition of the first kind (Dirichlet condition);

(2) boundary condition of the second kind (Neumann condition);

(3) boundary condition of the third kind (Robin condition).

For example, the boundary value problem with the condition of the first kind (the Dirichlet boundary value problem) for the Laplace equation is formulated as follows: find the solution of the equation $\Delta u = 0$ in some domain of space (or of the plane) which takes given values on the boundary. As a concrete physical example one can give the problem of determining the steady temperature distribution inside a domain Ω, if the temperature on its boundary $\partial\Omega$ is given. Another example: find the distribution of electric potential inside a domain if the potential on its boundary is known. The mathematical model of both phenomena is

$$\begin{cases} \Delta u = 0 & \text{in the domain } \Omega, \\ u = \varphi & \text{on the boundary } \partial\Omega, \end{cases}$$

where φ is a given function

The boundary value problem with boundary condition of the second kind (the Neumann boundary value problem) is posed as follows: find the solution of the given equation in some domain Ω of space (or of the plane) assuming that the outward normal derivative $\partial u/\partial n$ (which is proportional to the heat or mass flux) is given on $\partial\Omega$. This general boundary value problem, for the heat equation or for the equation of electrostatics, with the flux given on the boundary, is written as follows:

$$\begin{cases} \Delta u = 0 & \text{in the domain } \Omega, \\ \dfrac{\partial u}{\partial n} = \varphi & \text{on the boundary } \partial\Omega. \end{cases}$$

Unlike the Dirichlet problem for the Laplace equation, the Neumann problem has a meaning only in the case when the total flux through the boundary $\partial\Omega$ is equal to zero, i.e., $\int_{\partial\Omega} \frac{\partial u}{\partial n}\, ds = 0$. For example, the interior Neumann problem for the unit disc,

$$\begin{cases} \Delta u = 0 & 0 \le \rho < 1, \quad 0 \le \varphi \le 2\pi, \\ \dfrac{\partial u}{\partial \rho}(1,\varphi) = 1, & 0 \le \varphi \le 2\pi, \end{cases}$$

does not have a physical meaning because a constant unit flux inside the domain cannot ensure that the solution is stationary.

The Dirichlet and Neumann boundary value problems for the Poisson equation $\Delta u = f$ are formulated in a similar manner. Let us mention only that for the Neumann boundary value problem

$$\begin{cases} \Delta u = f & \text{in the domain } \Omega, \\ \dfrac{\partial u}{\partial n} = \varphi & \text{on the boundary } \partial\Omega, \end{cases}$$

to have a solution it is necessary and sufficient that

$$\int_\Omega f\, dx = \int_{\partial\Omega} \varphi\, ds.$$

Another peculiarity of the Neumann problem for the Poisson equation, which distinguishes it from the other boundary value problems considered here, is that its solution is not unique.

The boundary value problem with boundary condition of the third kind (the Robin boundary value problem) for, say, the Poisson equation, is posed as follows: find a solution $u(M)$ of the equation in some domain Ω of space (or of the plane), which satisfies on the boundary $\partial\Omega$ the condition $\partial u/\partial n + \sigma u = \varphi$, where σ and φ are given functions on $\partial\Omega$. More precisely, this problem reads

$$\begin{cases} \Delta u = f & \text{in the domain } \Omega, \\ \dfrac{\partial u}{\partial n} + \sigma u = \varphi & \text{on the boundary } \partial\Omega. \end{cases}$$

The solvability of this problem depends in essential manner on the behavior of the function σ, in particular, on its sign.

The aim of the present book is to consider a number of methods for solving problems of mathematical physics. The various methods are described as formal procedures, without any attempt to provide a mathematical justification.

The reader is not required to understand the physical essence of the examples that are used to illustrate the methods. However, familiarity with the basics of calculus, as well as with methods for solving ordinary differential equations is assumed.

Let us describe briefly the contents of this handbook.

Chapter 1 is devoted to elliptic problems. It treats the Fourier method for the Laplace and Poisson equation in domains with a certain symmetry, the Fourier method for the Helmholtz equation (in bounded as well as unbounded domains), and the Green function method.

In Chapter 2 a number of hyperbolic problems are studied. We consider the application of the methods of travelling and steady waves in one-dimensional as well as higher-dimensional cases. We demonstrate the application of the methods of the Fourier, Laplace and Hankel transformations, as also of the method of separation of variables to solving certain types of problems. therein we also illustrate the application of the perturbation method to some hyperbolic problems.

Chapter 3 is devoted to parabolic problems. It treats the application of the Fourier method in domains with a certain symmetry. The methods of the Fourier and Laplace integral transformations, as well as the method of separation of variables are applied to solve the Cauchy problem in the case of the homogeneous heat equation in one- and higher-dimensional situations.

Each chapter ends with a number of problems for independent study and answers to them.

To conclude let us mention that due to space limitations the book contains no material on systems of partial differential equations, probabilistic methods for solving boundary value problems, or variational methods.

Chapter 1
Elliptic problems

An effective method for solving boundary value problems for the Laplace and Helmoltz equations (in domains possessing a definite symmetry) is the *method of separation of variables*. The general idea of this method is to find a set of solutions of the homogeneous partial differential equation in question that satisfy certain boundary conditions. These solutions then serve as "atoms", from which, based on the linear superposition principle, one constructs the "general" solution. Since each of these "atoms" is a solution of the corresponding homogeneous equation, their linear combination is also a solution of the same equation. The solution of our problem is given by a series $\sum_{n=1}^{\infty} c_n u_n(x)$ (where $u_n(x)$ are the atom solutions, $x = (x_1, \dots, x_N)$ is the current point of the domain of space under consideration, and c_n are arbitrary constants). It remains to find constants c_n such that the boundary conditions are satisfied.

1.1. The Dirichlet problem for the Laplace equation in an annulus

Suppose that we are required to solve the Dirichlet problem for the Laplace equation $\Delta u = 0$ in the domain bounded by two concentric circles L_1 and L_2 centered at the origin, of radii R_1 and R_2:

$$\begin{cases} u_{xx} + u_{yy} = 0, & R_1^2 < x^2 + y^2 < R_2^2, \\ u|_{L_1} = f_1, & u|_{L_2} = f_2. \end{cases}$$

Introducing polar coordinates (ρ, φ), this Dirichlet problem can be recast as

$$\begin{cases} \rho^2 u_{\rho\rho} + \rho u_\rho + u_{\varphi\varphi} = 0, & R_1 < \rho < R_2, \quad 0 \le \varphi < 2\pi, \\ u(R_1, \varphi) = f_1(\varphi), \\ u(R_2, \varphi) = f_2(\varphi), & 0 \le \varphi < 2\pi. \end{cases} \tag{1.1}$$

The boundary functions $f_1(\varphi)$ and $f_2(\varphi)$ will be assumed to be 2π-periodic.

To solve the problem we will apply Fourier's method. Namely, we will seek the solution in the form $u(\rho, \varphi) = R(\rho)\Phi(\varphi)$. Substituting this expression in equation (1.1), we obtain

$$\Phi \rho^2 R'' + \Phi \rho R' + R\Phi'' = 0.$$

Next, dividing both sides of this equation by $R\Phi$ we get

$$\frac{\rho^2 R'' + \rho R'}{R} = -\frac{\Phi''}{\Phi}. \tag{1.2}$$

One says that in equation (1.2) the *variables are separated*, since the left-[resp., right-] hand side of the equation depends only on ρ [resp., φ]. Since the variables ρ and φ do not depend of one another, each of the two sides of equation (1.2) must be a constant. Let us denote this constant by λ. Then

$$\frac{\rho^2 R'' + \rho R'}{R} = -\frac{\Phi''}{\Phi} = \lambda. \tag{1.3}$$

It is clear that when the angle φ varies by 2π the single-valued function $u(\rho, \varphi)$ must return to the initial value, i.e., $u(\rho, \varphi) = u(\rho, \varphi + 2\pi)$. Consequently, $R(\rho)\Phi(\varphi) = R(\rho)\Phi(\varphi + 2\pi)$, whence $\Phi(\varphi) = \Phi(\varphi + 2\pi)$, i.e., the function $\Phi(\varphi)$ is 2π-periodic. From the equation $\Phi'' + \lambda\Phi = 0$ it follows that $\Phi(\varphi) = A\cos(\sqrt{\lambda}\varphi) + B\sin(\sqrt{\lambda}\varphi)$ (with A and B arbitrary constants), and in view of the periodicity of $\Phi(\varphi)$ we necessarily have $\lambda = n^2$, where $n \geq 0$ is an integer.

Indeed, the equality

$$A\cos(\sqrt{\lambda}\varphi) + B\sin(\sqrt{\lambda}\varphi) = A\cos[\sqrt{\lambda}(\varphi + 2\pi)] + B\sin[\sqrt{\lambda}(\varphi + 2\pi)]$$

implies that

$$\sin(\alpha + \sqrt{\lambda}\varphi) = \sin(\alpha + \sqrt{\lambda}\varphi + 2\pi\sqrt{\lambda}),$$

where we denote

$$\sin\alpha = \frac{A}{\sqrt{A^2 + B^2}}, \qquad \cos\alpha = \frac{B}{\sqrt{A^2 + B^2}}.$$

Therefore, $\sin(\pi\sqrt{\lambda})\cos(\alpha + \sqrt{\lambda}\varphi + \pi\sqrt{\lambda}) = 0$, i.e., $\pi\sqrt{\lambda} = \pi n$, or $\lambda = n^2$, where $n \geq 0$ is an integer. Now equation (1.3) yields

$$\rho^2 R'' + \rho R' - n^2 R = 0. \tag{1.4}$$

If $n \neq 0$, then we seek the solution of this equation in the form $R(\rho) = \rho^\mu$. Substituting this expression in equation (1.4) and simplifying by ρ^μ, we get

$$\mu^2 = n^2, \quad \text{or} \quad \mu = \pm n \quad (n > 0).$$

For $n = 0$ equation (1.4) has two solutions: 1 and $\ln\rho$. Thus, we now have an infinite set of functions ("atom" solutions)

$$1, \qquad \ln\rho, \qquad \rho^n\cos(n\varphi), \qquad \rho^n\sin(n\varphi),$$
$$\rho^{-n}\cos(n\varphi), \qquad \rho^{-n}\sin(n\varphi), \qquad n = 1, 2\ldots,$$

which satisfy the given partial differential equation. Since a sum of such solutions is also a solution, we conclude that in our case the "general" solution of the Laplace equation has the form

$$u(\rho, \varphi) = a_0 + b_0\ln\rho +$$
$$+ \sum_{n=1}^{\infty}\left[\left(a_n\rho^n + b_n\rho^{-n}\right)\cos(n\varphi) + \left(c_n\rho^n + d_n\rho^{-n}\right)\sin(n\varphi)\right]. \tag{1.5}$$

It remains only to find all the coefficients in the sum (1.5) so that the boundary conditions $u(R, \varphi) = f_1(\varphi)$, $u(R_2, \varphi) = f_2(\varphi)$ will be satisfied. Setting $\rho = R_1$ and then $\rho = R_2$ in (1.5) we obtain

$$u(R_1, \varphi) = \sum_{n=1}^{\infty} \left[\left(a_n R_1^n + b_n R_1^{-n} \right) \cos(n\varphi) + \right.$$
$$\left. + \left(c_n R_1^n + d_n R_1^{-n} \right) \sin(n\varphi) \right] + a_0 + b_0 \ln R_1,$$

$$u(R_2, \varphi) = \sum_{n=1}^{\infty} \left[\left(a_n R_2^n + b_n R_2^{-n} \right) \cos(n\varphi) + \right.$$
$$\left. + \left(c_n R_2^n + d_n R_2^{-n} \right) \sin(n\varphi) \right] + a_0 + b_0 \ln R_2.$$

Recalling the expressions for the Fourier coefficients of a trigonometric series, we arrive at the following systems of equations:

$$\begin{cases} a_0 + b_0 \ln R_1 = \dfrac{1}{2\pi} \displaystyle\int_0^{2\pi} f_1(s)\, ds, \\[2mm] a_0 + b_0 \ln R_2 = \dfrac{1}{2\pi} \displaystyle\int_0^{2\pi} f_2(s)\, ds, \end{cases} \tag{1.6_1}$$

(to be solved for a_0 and b_0);

$$\begin{cases} a_n R_1^n + b_n R_1^{-n} = \dfrac{1}{\pi} \displaystyle\int_0^{2\pi} f_1(s) \cos(ns)\, ds, \\[2mm] a_n R_2^n + b_n R_2^{-n} = \dfrac{1}{\pi} \displaystyle\int_0^{2\pi} f_2(s) \cos(ns)\, ds, \end{cases} \tag{1.6_2}$$

(to be solved for a_n and b_n); and

$$\begin{cases} c_n R_1^n + d_n R_1^{-n} = \dfrac{1}{\pi} \displaystyle\int_0^{2\pi} f_1(s) \sin(ns)\, ds, \\[2mm] c_n R_2^n + b d_n R_2^{-n} = \dfrac{1}{\pi} \displaystyle\int_0^{2\pi} f_2(s) \sin(ns)\, ds, \end{cases} \tag{1.6_3}$$

(to be solved for c_n and d_n).

Thus, from these systems one can find all the unknown coefficients a_0, b_0, a_n, b_n, c_n, d_n. Now the problem (1.1) is completely solved. The solution is given by the expression (1.5), in which the coefficients are obtained from the systems (1.6).

1.2. Examples of Dirichlet problem in an annulus

Example 1. Let us assume that the potential is equal to zero on the inner circle, and is equal to $\cos\varphi$ on the outer circle. Find the potential in the annulus.

We have to solve the problem

$$\begin{cases} \Delta u = 0, & 1 < \rho < 2, \quad 0 \le \varphi < 2\pi, \\ u(1,\varphi) = 0, & u(2,\varphi) = \cos\varphi, \quad 0 \le \varphi \le 2\pi, \end{cases}$$

in order to find determine the potential $u(\rho,\varphi)$ in the annulus.

Generally speaking, to solve this problem we have to calculate all the integrals in the formulas (1.6), and then solve the corresponding systems of equations to find the coefficients a_0, b_0, a_n, b_n, c_n, d_n. However, in the present case it is simpler to try to choose particular solutions such that a linear combination of them will satisfy the boundary conditions. Here such a role is played by the linear combination $u(\rho,\varphi) = a_1\rho\cos\varphi + b_1\rho^{-1}\cos\varphi$. The boundary conditions yield the system of equations

$$\begin{cases} a_1 + b_1 = 0, \\ 2a_1 + \dfrac{b_1}{2} = 1, \end{cases}$$

from which we find $a_1 = 2/3$, $b_1 = -2/3$. Therefore, the solution is

$$u(\rho,\theta) = \frac{2}{3}\left(\rho - \rho^{-1}\right)\cos\varphi.$$

Example 2. Let us consider the following problem with constant potentials on the boundaries of the annulus:

$$\begin{cases} \Delta u = 0, & 1 < \rho < 2, \quad 0 \le \varphi < 2\pi, \\ u(1,\varphi) = 2, & u(2,\varphi) = 1, \quad 0 \le \varphi < 2\pi. \end{cases}$$

In this case we will seek the solution as a function that does not depend on φ, i.e., $u(\rho) = a_0 + b_0\ln\rho$. Substituting this expression in the boundary conditions we obtain the system of equations

$$\begin{cases} a_0 + b_0\ln 1 = 2, \\ a_0 + b_0\ln 2 = 1, \end{cases}$$

which yields $a_0 = 2$, $b_0 = -\log_2 e$. Therefore, the sought solution is the function

$$u(\rho) = 2 - \frac{\ln\rho}{\ln 2}.$$

Example 3. Let us solve the following Dirichlet problem

$$\begin{cases} \Delta u = 0, & 1 < \rho < 2, \quad 0 \le \varphi < 2\pi, \\ u(1,\varphi) = \cos\varphi, & u(2,\varphi) = \sin\varphi, \quad 0 \le \varphi \le 2\pi. \end{cases}$$

One can verify that here all the coefficients a_0, b_0, a_n, b_n, c_n, d_n with $n > 1$ are equal to zero, while the coefficients a_1, b_1, c_1, d_1 are determined from the systems of equations

$$\begin{cases} a_1 + b_1 = 1, \\ 2a_1 + \dfrac{b_1}{2} = 0, \end{cases} \qquad \begin{cases} c_1 + d_1 = 0, \\ 2c_1 + \dfrac{d_1}{2} = 1. \end{cases}$$

Solving these systems we obtain

$$a_1 = -\frac{1}{3}, \qquad b_1 = \frac{4}{3}, \qquad c_1 = \frac{2}{3}, \qquad d_1 = -\frac{2}{3}.$$

Thus, the solution of our problem is the function

$$u(\rho, \varphi) = \left(-\frac{1}{3}\rho + \frac{4}{3\rho} \right) \cos\varphi + \frac{2}{3}\left(\rho - \frac{1}{\rho} \right) \sin\varphi.$$

Since the Dirichlet problem for the Laplace equation in a bounded domain has a unique solution, in examples 1–3 there are no other solutions besides the ones found.

1.3. The interior and exterior Dirichlet problems

Let us consider the two very important cases in which the annulus becomes a disc or the exterior of a disc. The *interior* Dirichlet problem ($R_1 = 0$, $R_2 = R$)

$$\begin{cases} \rho^2 u_{\rho\rho} + \rho u_\rho + u_{\varphi\varphi} = 0, & 0 \le \rho < R, \quad 0 \le \varphi < 2\pi, \\ u(R, \varphi) = f(\varphi), & 0 \le \varphi \le 2\pi, \end{cases}$$

is solved in exactly the same manner as the Dirichlet problem for the annulus, with the only difference that now we must discard the solution "atoms" that are not bounded when ρ approaches 0:

$$\ln\rho, \qquad \rho^{-n}\cos(n\varphi), \qquad \rho^{-n}\sin(n\varphi), \qquad n = 1, 2, \ldots$$

Hence, the solution is given by the remaining terms, i.e.,

$$u(\rho, \varphi) = \sum_{n=0}^{\infty} \left(\frac{\rho}{R} \right)^n [a_n \cos(n\varphi) + b_n \sin(n\varphi)],$$

where the coefficients a_n and b_n are calculated by means of the formulas

$$\left. \begin{aligned} a_0 &= \frac{1}{2\pi} \int_0^{2\pi} f(\varphi)\,d\varphi, \\ a_n &= \frac{1}{\pi} \int_0^{2\pi} f(\varphi)\cos(n\varphi)\,d\varphi, & n > 0, \\ b_n &= \frac{1}{\pi} \int_0^{2\pi} f(\varphi)\sin(n\varphi)\,d\varphi, & n > 0. \end{aligned} \right\} \tag{1.7}$$

In other words, we simply expand the function $f(\varphi)$ in a Fourier series

$$f(\varphi) = \sum_{n=0}^{\infty} [a_n \cos(n\varphi) + b_n \sin(n\varphi)],$$

and then multiply each term of the series by the factor $\left(\frac{\rho}{R}\right)^n$. For example, the interior problem

$$\begin{cases} \Delta u = 0, & 0 \le \rho < 1, \quad 0 \le \varphi < 2\pi, \\ u(1, \varphi) = \cos^2 \varphi, & 0 \le \varphi < 2\pi, \end{cases}$$

has the solution

$$u(\rho, \varphi) = \frac{1}{2} + \frac{1}{2} \rho^2 \cos(2\varphi).$$

The *exterior* Dirichlet problem ($R_1 = R$, $R_2 = \infty$)

$$\begin{cases} \rho^2 u_{\rho\rho} + \rho u_\rho + u_{\varphi\varphi} = 0, & R \le \rho < \infty, \quad 0 \le \varphi < 2\pi, \\ u(R, \varphi) = f(\varphi), & 0 \le \varphi \le 2\pi, \end{cases}$$

is solved in much the same way as the preceding problem, with the difference than now we discard the solution "atoms" that are not bounded when ρ goes to infinity:

$$\ln \rho, \qquad \rho^n \cos(n\varphi), \qquad \rho^n \sin(n\varphi), \qquad n = 1, 2, \ldots$$

Accordingly, the solution is taken in the form

$$u(\rho, \varphi) = \sum_{n=0}^{\infty} \left(\frac{\rho}{R}\right)^{-n} [a_n \cos(n\varphi) + b_n \sin(n\varphi)],$$

where the coefficients a_n and b_n are calculated by means of formulas (1.7). For example, the exterior problem

$$\begin{cases} \Delta u = 0, & 1 \le \rho < \infty, \quad 0 \le \varphi < 2\pi, \\ u(1, \varphi) = \sin^3 \varphi, & 0 \le \varphi \le 2\pi \end{cases}$$

has the solution

$$u(\rho, \varphi) = \frac{3}{4} \cdot \frac{1}{\rho} \sin \varphi - \frac{1}{4} \cdot \frac{1}{\rho^3} \sin 3\varphi.$$

Let us note that the Dirichlet problem for the Laplace equations in an unbounded two-dimensional domain has only one bounded solution.

We conclude this section by examining another example, one exercise (the Poisson integral), and a problem connected with the Poisson integral.

Example [1]. Find the steady temperature distribution in a homogeneous sector $0 \le \rho \le a$, $0 \le \varphi \le \alpha$, which satisfies the boundary conditions $u(\rho, 0) = u(\rho, \alpha) = 0$, $u(a, \varphi) = A\varphi$, where A is a constant (see Figure 1.1).

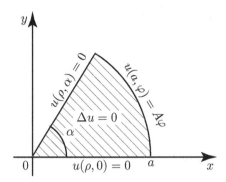

FIGURE 1.1.

Solution. Finding the steady temperature distribution reduces to solving the Dirichlet problem

$$\begin{cases} \rho^2 u_{\rho\rho} + \rho u_\rho + u_{\varphi\varphi} = 0, & 0 \le \rho < a, \quad 0 < \varphi < \alpha < 2\pi, \\ u(\rho, 0) = u(\rho, \alpha) = 0, & 0 \le \rho \le a, \\ u(a, \varphi) = A\varphi, & 0 \le \varphi \le \alpha. \end{cases}$$

Setting $u(\rho, \varphi) = R(\rho)\Phi(\varphi)$ and separating variables, we obtain two ordinary differential equations:

$$\begin{aligned} \rho^2 R'' + \rho R' - \lambda R = 0, \\ \Phi'' + \lambda\Phi = 0. \end{aligned} \tag{1.8}$$

The conditions $0 = u(\rho, 0) = R(\rho)\Phi(0)$ and $0 = u(\rho, \alpha) = R(\rho)\Phi(\alpha)$ yield $\Phi(0) = \Phi(\alpha) = 0$. The separation constant λ is determined by solving the Sturm-Liouville

$$\begin{cases} \Phi'' + \lambda\Phi = 0, & 0 < \varphi < \alpha, \\ \Phi(0) = \Phi(\alpha) = 0. \end{cases}$$

We get $\lambda_n = \left(\frac{n\pi}{\alpha}\right)^2$ and

$$\mu(\mu - 1) + \mu - \left(\frac{n\pi}{\alpha}\right)^2 = 0,$$

whence

$$\mu = \pm \frac{n\pi}{\alpha}.$$

Using the fact that the function $R(\rho)$ is bounded (according to the meaning of the problem at hand), we write $R_n(\rho) = \rho^{n\pi/\alpha}$. The atoms from which our solution is built are the functions

$$u_n(\rho, \varphi) = \rho^{n\pi/\alpha} \sin\left(\frac{n\pi}{\alpha}\varphi\right), \qquad n = 1, 2, \ldots$$

Thus, the solution itself is

$$u(\rho, \varphi) = \sum_{n=1}^{\infty} c_n \rho^{n\pi/\alpha} \sin\left(\frac{n\pi}{\alpha}\varphi\right).$$

The constants c_n $(n = 1, 2, \ldots)$ are found from the condition $u(a, \varphi) = A\varphi$. Since

$$u(a, \varphi) = \sum_{n=1}^{\infty} c_n a^{n\pi/\alpha} \sin\left(\frac{n\pi}{\alpha}\varphi\right).$$

it follows that

$$c_n a^{n\pi/\alpha} = \frac{2}{\alpha} \int_0^\alpha A\varphi \sin\left(\frac{n\pi}{\alpha}\varphi\right) d\varphi,$$

and so

$$c_n = \frac{2A}{\alpha a^{n\pi/\alpha}} \int_0^\alpha \varphi \sin\left(\frac{n\pi}{\alpha}\varphi\right) d\varphi = (-1)^{n+1}\frac{2\alpha A}{n\pi}.$$

Finally, the solution of our problem is written in the form

$$u(\rho, \varphi) = \frac{2\alpha A}{\pi} \sum_{n=1}^{\infty} (-1)^{n+1} \left(\frac{\rho}{\alpha}\right)^{n\pi/\alpha} \frac{\sin\left(\frac{n\pi}{\alpha}\right)\varphi}{n}.$$

Notice that the solution has a singularity in the boundary point $\rho = a$, $\varphi = \alpha$ because of the incompatibility of the boundary values.

1.4. The Poisson integral for the disc. Complex form. Solution of the Dirichlet problem when the boundary condition is a rational function $R(\sin\varphi, \cos\varphi)$

Recall that the solution of the interior and exterior Dirichlet problem is can be presented in integral form (the Poisson integral):

$$u(\rho, \varphi) = \frac{1}{2\pi} \int_0^{2\pi} \frac{R^2 - \rho^2}{R^2 - 2\rho R \cos(\varphi - \alpha) + \rho^2} f(\alpha)\, d\alpha, \quad \rho < R,$$

$$u(\rho, \varphi) = \frac{1}{2\pi} \int_0^{2\pi} \frac{\rho^2 - R^2}{R^2 - 2\rho R \cos(\varphi - \alpha) + \rho^2} f(\alpha)\, d\alpha, \quad \rho > R.$$

Let us show that these formulas are a consequence of the general superposition method.

For the sake of definiteness we shall consider the interior problem, and then write the result for the exterior problem by analogy.

Substituting the expression for the Fourier coefficients in the formula

$$u(\rho, \varphi) = \sum_{n=0}^{\infty} \left(\frac{\rho}{R}\right)^n [a_n \cos(n\varphi) + b_n \sin(n\varphi)],$$

we obtain

$$u(\rho, \varphi) = \frac{1}{\pi} \int_0^{2\pi} f(\alpha) \left[\frac{1}{2} + \sum_{n=0}^{\infty} \left(\frac{\rho}{R}\right)^n (\cos(n\varphi)\cos(n\alpha) + \sin(n\varphi)\sin(n\alpha))\right] =$$

$$= \frac{1}{\pi} \int_0^{2\pi} f(\alpha) \left[\frac{1}{2} + \sum_{n=0}^{\infty} \left(\frac{\rho}{R}\right)^n \cos(n(\varphi - \alpha))\right] d\alpha.$$

Further, using the relation $\cos(n(\varphi - \alpha)) = \frac{1}{2}\left(e^{in(\varphi-\alpha)} + e^{-in(\varphi-\alpha)}\right)$, the fact that $q = \rho/R < 1$ and the formula for the sum of an infinite decreasing geometric progresion, we get

$$\frac{1}{2} + \sum_{n=1}^{\infty} q^n \cos(n(\varphi - \alpha)) = \frac{1}{2} + \frac{1}{2}\sum_{n=1}^{\infty} q^n \left[e^{in(\varphi-\alpha)} + e^{-in(\varphi-\alpha)}\right] =$$

$$= \frac{1}{2}\left[1 + \sum_{n=1}^{\infty}\left[\left(qe^{in(\varphi-\alpha)}\right)^n + \left(qe^{-in(\varphi-\alpha)}\right)^n\right]\right] =$$

$$= \frac{1}{2}\left[1 + \frac{qe^{in(\varphi-\alpha)}}{1 - qe^{in(\varphi-\alpha)}} + \frac{qe^{-in(\varphi-\alpha)}}{1 - qe^{-in(\varphi-\alpha)}}\right] =$$

$$= \frac{1}{2} \cdot \frac{1 - q^2}{1 - 2q\cos(\varphi - \alpha) + q^2} = \frac{1}{2} \cdot \frac{R^2 - \rho^2}{R^2 - 2R\rho\cos(\varphi - \alpha) + \rho^2}.$$

Therefore,

$$u(\rho, \varphi) = \frac{1}{2\pi} \int_0^{2\pi} \frac{R^2 - \rho^2}{R^2 - 2R\rho\cos(\varphi - \alpha) + \rho^2} f(\alpha)\, d\alpha, \qquad \rho < R.$$

Let us recast the Poisson formula in a different form (complex notation). Note that

$$\frac{R^2 - \rho^2}{R^2 - 2R\rho\cos(\varphi - \alpha) + \rho^2} = \frac{R^2 - |z|^2}{|Re^{i\alpha} - z|^2} = \mathrm{Re}\frac{Re^{i\alpha} + z}{Re^{i\alpha} - z}$$

because

$$\operatorname{Re} \frac{Re^{i\alpha} + z}{Re^{i\alpha} - z} = \operatorname{Re} \frac{\left(Re^{i\alpha} + \rho e^{i\varphi}\right)\left(\overline{Re^{i\alpha}} - \overline{\rho e^{i\varphi}}\right)}{\left(Re^{i\alpha} - \rho e^{i\varphi}\right)\left(\overline{Re^{i\alpha}} - \overline{\rho e^{i\varphi}}\right)}$$

$$= \operatorname{Re} \frac{R^2 - |z|^2 + \rho R\left[e^{i(\varphi-\alpha)} - e^{i(\varphi-\alpha)}\right]}{|Re^{i\alpha} - z|^2} = \frac{R^2 - |z|^2}{|Re^{i\alpha} - z|^2}.$$

It follows that the Poisson integral can be written in the form

$$u(z) = \operatorname{Re} \frac{1}{2\pi} \int_0^{2\pi} \frac{Re^{i\alpha} + z}{Re^{i\alpha} - z} f(\alpha)\, d\alpha.$$

If in this integral we set $\zeta = Re^{i\alpha}$ and, accordingly, $d\alpha = d\zeta/i\zeta$, we finally obtain

$$u(z) = \operatorname{Re} \frac{1}{2\pi i} \int_0^{2\pi} \frac{\zeta + z}{\zeta - z} f(\zeta) \frac{d\zeta}{\zeta}, \qquad |z| < R. \tag{1.9}$$

If the boundary function $f(\zeta)$ is a rational function of $\sin\varphi$ and $\cos\varphi$, then the integral in formula (1.9) can be calculated by means of residues.

Example. Solve the Dirichlet problem

$$\begin{cases} \Delta u = 0, & |z| < 2, \\ u|_{|z|=2} = \dfrac{2\sin\varphi}{5 + 3\cos\varphi}. \end{cases}$$

Solution. We shall use formula (1.9). Let $\zeta = 2e^{i\alpha}$; then

$$\sin\alpha = \frac{1}{2i}\left(\frac{\zeta}{2} - \frac{2}{\zeta}\right) \qquad \cos\alpha = \frac{1}{2}\left(\frac{\zeta}{2} + \frac{2}{\zeta}\right)$$

and the boundary function becomes

$$u(\zeta) = \frac{2\sin\alpha}{5 + 3\cos\alpha} = \frac{2 \cdot \dfrac{1}{2i} \cdot \dfrac{\zeta^2 - 4}{2\zeta}}{5 + \dfrac{3}{2}\left(\dfrac{\zeta}{2} + \dfrac{2}{\zeta}\right)} =$$

$$= \frac{2}{i} \cdot \frac{\zeta^2 - 4}{3\zeta^2 + 20\zeta + 12} = \frac{2}{i} \cdot \frac{\zeta^2 - 4}{3(\zeta + 6)\left(\zeta + \dfrac{2}{3}\right)}.$$

Let us compute the integral

$$J = \frac{1}{2\pi i} \int_{|\zeta|=2} \frac{2(\zeta^2 - 4)(\zeta + z)}{i \cdot 3(\zeta + 6)(\zeta + \dfrac{2}{3})(\zeta - z)\zeta}\, d\zeta$$

where the circle $|\zeta| = 2$ is oriented counter-clockwise. In our case the integrand $F(\zeta)$ has in the domain $|\zeta| > 2$ only one finite singular point $\zeta = -6$ – a pole of order one – and the removable singular point $\zeta = \infty$. By the Cauchy residue theorem,

$$J = -\operatorname{res}[F(\zeta)]_{\zeta=-6} - \operatorname{res}[F(\zeta)]_{\zeta=\infty}.$$

First let us find the residue at the point $\zeta = -6$;

$$\operatorname{res}[F(\zeta)]_{\zeta=-6} = \frac{2}{3i} \cdot \frac{32}{\left(-\frac{16}{3}\right)} \cdot \frac{z-6}{(z+6)\cdot 6} = -\frac{4}{i} \cdot \frac{z-6}{(z+6)\cdot 6} = \frac{2}{3i} \cdot \frac{6-z}{6+z}.$$

Next let us expand $F(\zeta)$ in a series in the neighborhood of the point $\zeta = \infty$:

$$F(\zeta) = \frac{2}{3i} \cdot \frac{\left(1 - \frac{4}{\zeta^2}\right)\left(1 + \frac{z}{\zeta}\right)}{\left(1 + \frac{6}{\zeta}\right)\left(1 + \frac{2}{3\zeta}\right)} \cdot \frac{1}{1 - \frac{z}{\zeta}} \cdot \frac{1}{\zeta} = \frac{2}{3i} \cdot \frac{1}{\zeta} + \dots ,$$

whence

$$\operatorname{res}[F(\zeta)]_{\zeta=\infty} = -\frac{2}{3i}.$$

Therefore,

$$J = \frac{2}{3i} \cdot \frac{z-6}{z+6} + \frac{2}{3i} = \frac{2}{3i} \cdot \frac{2z}{z+6} = \frac{4z}{3i(z+6)} =$$

$$= \frac{4}{3i} \cdot \frac{x+iy}{6+x+iy} = \frac{4}{3i} \cdot \frac{(x+iy)(6+x-iy)}{(6+x)^2 + y^2},$$

which yields

$$\operatorname{Re} J = \frac{8y}{36 + 12x + x^2 + y^2},$$

or

$$\operatorname{Re} J = \frac{8\rho \sin \varphi}{36 + 12\rho \cos \varphi + \rho^2}.$$

We conclude that the solution of our Dirichlet problem is given by the expression

$$u(\rho, \varphi) = \frac{8\rho \sin \varphi}{36 + 12\rho \cos \varphi + \rho^2}.$$

1.5. The interior and exterior Neumann problems for a disc

It is clear that in the case of a disc of radius R centered at the origin the exterior normal derivative is $\partial u/\partial n|_{\rho=R} = \partial u/\partial\rho|_{\rho=R}$. Accordingly, the solution of the interior Neumann problem is sought in the form of a series

$$u(\rho, \varphi) = \sum_{n=0}^{\infty} \left(\frac{\rho}{R}\right)^n [a_n \cos(n\varphi) + b_n \sin(n\varphi)].$$

The coefficients a_n and b_n of this series are determined from the boundary condition $\partial u/\partial\rho|_{\rho=R} = f(\varphi)$, i.e., we have

$$\begin{aligned}
a_n &= \frac{R}{n\pi} \int_0^{2\pi} f(\varphi) \cos(n\varphi)\, d\varphi, \\
b_n &= \frac{R}{n\pi} \int_0^{2\pi} f(\varphi) \sin(n\varphi)\, d\varphi,
\end{aligned} \qquad n = 1, 2, \ldots \qquad (1.10)$$

Similarly, the solution of the exterior Neumann problem is sought in the form of a series

$$u(\rho, \varphi) = \sum_{n=0}^{\infty} \left(\frac{\rho}{R}\right)^{-n} [a_n \cos(n\varphi) + b_n \sin(n\varphi)].$$

whose coefficients a_n and b_n, determined from the boundary condition $\partial u/\partial\rho|_{\rho=R} = f(\varphi)$, are calculated by means of the same formulas (1.10) (here we use the fact that $\partial u/\partial n|_{\rho=R} = -\partial u/\partial\rho|_{\rho=R}$).

Example. Find the steady temperature inside of an unbounded cylinder of radius R if on the lateral surface S there is given the heat flux $\partial u/\partial n|_S = \cos^3 \varphi$.

Solution. We have to solve the interior Neumann problem

$$\begin{cases}
\Delta u = 0, & 0 < \rho < R, \quad 0 \leq \varphi < 2\pi, \\
\left.\dfrac{\partial u}{\partial \rho}\right|_{\rho=R} = \cos^3 \varphi, & 0 \leq \varphi \leq 2\pi.
\end{cases}$$

First of all we need to verify that the condition for the solvability of the Neumann problem is satisfied, i.e., that $\int_C \frac{\partial u}{\partial n}\, ds = 0$, where C is the circle bounding our disc.

Indeed, we have

$$\int_C \frac{\partial u}{\partial n}\, ds = \int_0^{2\pi} \cos^3 \varphi \cdot R\, d\varphi =$$

$$= \frac{R}{2} \int_0^{2\pi} \cos \varphi\, d\varphi + \frac{R}{4} \int_0^{2\pi} [\cos(3\varphi) + \cos \varphi]\, d\varphi = 0.$$

Next, since $\cos^3 \varphi = \frac{3}{4} \cos \varphi + \frac{1}{4} \cos(3\varphi)$, it follows that $a_1 = \frac{3}{4} R$, $a_3 = \frac{1}{12} R$, and all the remaining coefficients in the series giving the solution of the interior Neumann problem are equal to zero. Hence, the solution has the form

$$u(\rho, \varphi) = C + \frac{3\rho}{4} \cos \varphi + \frac{\rho^3}{12 R^2} \cos(3\varphi),$$

where C is an arbitrary constant.

Remark. The Neumann problem can also be solved for an annulus. In this case the boundary conditions specify the exterior normal derivative:

$$-\frac{\partial u}{\partial \rho}(R_1, \varphi) = f_1(\varphi), \qquad \frac{\partial u}{\partial \rho}(R_2, \varphi) = f_2(\varphi).$$

Here the solution exists only if the condition

$$R_1 \int_0^{2\pi} -f_1(\varphi)\, d\varphi = R_2 \int_0^{2\pi} f_2(\varphi)\, d\varphi$$

is satisfied, and is uniquely determined up to an arbitrary constant.

1.6. Boundary value problems for the Poisson equation in a disc and in an annulus

When we solve the Dirichlet or Neumann problem (or a problem of mixed type) we need first to find some particular solution u_1 of the Poisson equation $\Delta u = f(x, y)$ and then use the change of dependent variables $u = u_1 + v$ to reduce the task to that of solving the corresponding boundary value problem for the Laplace equation $\Delta v = 0$.

Example 1 [18]. Solve the Poisson equation

$$\frac{\partial^2 u}{\partial x^2} + \frac{\partial^2 u}{\partial y^2} = -xy$$

in the disc of radius R centered at the origin, under the condition $u(R, \varphi) = 0$.

Solution. Passing to polar coordinates we obtain the problem

$$\begin{cases} \rho^2 u_{\rho\rho} + \rho u_\rho + u_{\varphi\varphi} = -\dfrac{1}{2} \rho^4 \sin(2\varphi), & 0 \le \rho < R, \quad 0 \le \varphi < 2\pi, \\ u(R, \varphi) = 0, & 0 \le \varphi \le 2\pi. \end{cases} \tag{1.11}$$

We shall seek a particular solution in the form

$$u_1(\rho, \varphi) = w(\rho) \sin(2\varphi).$$

Substituting this expression in equation (1.11) and simplifying by $\sin(2\varphi)$ we obtain the equation

$$\rho^2 w'' + \rho w' - 4w = -\frac{1}{2}\rho^4. \tag{1.12}$$

The substitution $\rho = e^t$ transforms (1.12) into the equation with constant coefficients

$$\ddot{w} - 4w = -\frac{1}{2}e^{4t}, \tag{1.13}$$

where the dot denotes differrentiation with respect to t. A particular solution of equation (1.13) is $w(t) = -\frac{1}{24}e^{4t}$. Hence, $w(\rho) = -\frac{1}{24}\rho^4$ is a particular solution of equation (1.12). Therefore, we can choose $u_1(\rho, \varphi) = -\frac{1}{24}\rho^4 \sin(2\varphi)$.

Now let us introduce the function $v(\rho, \varphi) = u(\rho, \varphi) - u_1(\rho, \varphi)$. Clearly, to determine the function $v(\rho, \varphi)$ we must solve the following Dirichlet problem for the Laplace equation:

$$\begin{cases} \rho^2 u_{\rho\rho} + \rho v_\rho + v_{\varphi\varphi} = 0, & 0 < \rho < R, \quad 0 \le \varphi < 2\pi, \\ v(R, \varphi) = \dfrac{1}{24} R^4 \sin(2\varphi), & 0 \le \varphi \le 2\pi. \end{cases}$$

But we already know the solution of this equation:

$$v(\rho, \varphi) = \left(\frac{\rho}{R}\right)^2 \cdot \frac{1}{24} R^4 \sin(2\varphi) = \frac{1}{24}\rho^2 R^4 \sin(2\varphi).$$

Therefore, the solution of our problem is given by

$$u(\rho, \varphi) = \frac{1}{24}\rho^2(R^4 - \rho^2)\sin(2\varphi).$$

Example 2. Find the distribution of the electric potential in the annulus $a < \rho < b$ if in its interior there are electrical charges with density $\gamma(x, y) = A(x^2 - y^2)$, the inner circle is maintained at the potential 1 and the intensity of the electric field on the outer circle is 0.

Solution. The problem reduces to that of solving the Poisson equation $\Delta u = A(x^2 - y^2)$ in the annulus $a < \rho < b$ with the boundary conditions $u|_{\rho=a} = 1$, $\partial u / \partial \rho|_{\rho=b} = 0$. Passing to polar coordinates we obtain the problem

$$\begin{cases} \dfrac{1}{\rho}\dfrac{\partial}{\partial\rho}\left(\rho\dfrac{\partial u}{\partial\rho}\right) + \dfrac{1}{\rho^2}\dfrac{\partial^2 u}{\partial\varphi^2} = A\rho^2\cos(2\varphi) & a < \rho < b, \quad 0 \le \varphi < 2\pi, \\ u(a, \varphi) = 1, \quad \dfrac{\partial u}{\partial\rho}(b, \varphi) = 0, & 0 \le \varphi \le 2\pi. \end{cases}$$

Let us seek the solution of this problem in the form $u(\rho, \varphi) = v(\rho, \varphi) + w(\rho)$, where the function $w(\rho)$ is a solution of the auxiliary problem

$$\begin{cases} \dfrac{1}{\rho}\dfrac{\partial}{\partial\rho}\left(\rho\dfrac{\partial w}{\partial\rho}\right) = 0, & a < \rho < b, \\ w(a) = 1, \quad w'(b) = 0, \end{cases} \tag{1.14}$$

and the function $v(\rho, \varphi)$ is a solution of the problem

$$\begin{cases} \dfrac{1}{\rho}\dfrac{\partial}{\partial\rho}\left(\rho\dfrac{\partial v}{\partial\rho}\right) + \dfrac{1}{\rho^2}\dfrac{\partial^2 v}{\partial\varphi^2} = A\rho^2\cos(2\varphi) & a < \rho < b, \quad 0 \le \varphi < 2\pi, \\[3mm] v(a, \varphi) = 0, \quad \dfrac{\partial v}{\partial\rho}(b, \varphi) = 0, & 0 \le \varphi \le 2\pi. \end{cases} \tag{1.15}$$

Obviously, the solution of problem (1.14) is $w(\rho) \equiv 1$. We will seek the solution of problem (1.15) in the form $v(\rho, \varphi) = R(\rho)\cos(2\varphi)$. Substituting this expression for $v(\rho, \varphi)$ in the equation (1.15) we obtain

$$\cos(2\varphi)\frac{1}{\rho}\frac{d}{d\rho}(\rho R') - \frac{4}{\rho^4}R\cos(2\varphi) = A\rho^2\cos(2\varphi),$$

or, simplifying by $\cos(2\varphi)$,

$$\rho^2 R'' + \rho R' - 4R = A\rho^4,$$

with the additional conditions $R(a) = 0$, $R'(b) = 0$. The substitution $\rho = e^t$ transform this equation into the equation with constant coefficients

$$\ddot{R} - 4R = Ae^4 t,$$

where the dot denotes differentiation with respect to t. The general solution of this last equation is $R(t) = C_1 e^{2t} + C_2 e^{-2t} + \frac{1}{12}Ae^{4t}$. Back to the variable ρ we have

$$R(\rho) = C_1\rho^2 + \frac{C_2}{\rho^2} + \frac{1}{12}Ae^{4t}.$$

The constants C_1 and C_2 are found from the conditions $R(a) = 0$, $R'(b) = 0$, namely

$$C_1 = \frac{-A(a^6 + 2b^6)}{12(a^4 + b^4)}, \qquad C_2 = \frac{Aa^4 b^4(2b^2 - a^2)}{6(a^4 + b^4)}$$

Hence, the sought solution is

$$u(\rho, \varphi) = 1 + \left[-\frac{A(a^6 + 2b^6)}{12(a^4 + b^4)}\rho^2 + \frac{Aa^4 b^4(2b^2 - a^2)}{6(a^4 + b^4)}\frac{1}{\rho^2} + \frac{A}{12}\rho^4\right]\cos(2\varphi).$$

1.7. Boundary value problems for the Laplace and Poisson equations in a rectangle

Example 1 [18]. Find the distribution of the electrostatic field $u(x, y)$ inside the rectangle $OACB$ for which the potential along the side B is equal to V, while the three other sides and grounded. There are no electric charges inside the rectangle (Figure 1.2).

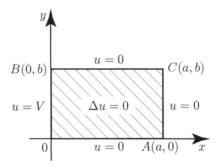

FIGURE 1.2.

Solution. The problem reduces to that of solving the Laplace equation $u_{xx} + u_{yy} = 0$ in the interior of the rectangle with the boundary conditions

$$u(0, y) = V, \quad u(a, y) = 0, \quad u(x, 0) = 0, \quad u(x, b) = 0.$$

First we will seek nontrivial particular solutions of the Laplace equation which satisfy only the boundary conditions

$$u(x, 0) = u(x, b) = b$$

in the form $u(x, y) = X(x)Y(y)$. Substituting this expression in the equation $u_{xx} + u_{yy} = 0$ we get $X''Y + XY'' = 0$, which upon dividing by XY gives

$$\frac{X''}{X} = -\frac{Y''}{Y} = \lambda^2.$$

Using the fact that $Y(0) = Y(b) = 0$, we obtain the Sturm-Liouville problem

$$\begin{cases} Y'' + \lambda^2 Y = 0, & 0 < y < b, \\ Y(0) = Y(b) = 0, \end{cases}$$

which yields the eigenvalues and eigenfunctions of our problem. We have

$$\lambda_n^2 = \left(\frac{n\pi}{b}\right)^2, \qquad Y_n(y) = \sin\left(\frac{n\pi}{b} y\right), \qquad n = 1, 2, \dots$$

The corresponding functions $X_n(x)$ are solutions of the equation $X'' - \lambda^2 X = 0$, and so
$$X_n(x) = a_n \cosh\left(\frac{n\pi}{b} x\right) + b_n \sinh\left(\frac{n\pi}{b} x\right)$$
where a_n and b_n are arbitrary constants. It follows that the notrivial particular solutions ("atoms") have the form
$$u_n(x,y) = \left[a_n \cosh\left(\frac{n\pi}{b} x\right) + b_n \sinh\left(\frac{n\pi}{b} x\right)\right] \sin\left(\frac{n\pi}{b} y\right), \qquad n = 1,2,\ldots$$

Now for the sought solution of our problem we take the series
$$u(x,y) = \sum_{n=0}^{\infty} \left[a_n \cosh\left(\frac{n\pi}{b} x\right) + b_n \sinh\left(\frac{n\pi}{b} x\right)\right] \sin\left(\frac{n\pi}{b} y\right). \qquad (1.16)$$

The constants a_n and b_n ($n = 1,2,\ldots$) are found from the conditions $u(0,y) = V$, $u(a,y) = 0$. Setting $x = a$ in (1.16) we obtain
$$0 = \sum_{n=0}^{\infty} \left[a_n \cosh\left(\frac{n\pi}{b} a\right) + b_n \sinh\left(\frac{n\pi}{b} a\right)\right] \sin\left(\frac{n\pi}{b} y\right),$$
whence
$$a_n \cosh\left(\frac{n\pi}{b} a\right) + b_n \sinh\left(\frac{n\pi}{b} a\right) = 0, \qquad n = 1,2,\ldots$$

Next, setting $x = 0$ in (1.16) we obtain
$$V = \sum_{n=0}^{\infty} a_n \sin\left(\frac{n\pi}{b} y\right),$$
which gives
$$a_n = \frac{2}{b} \int_0^b V \sin\left(\frac{n\pi}{b} y\right) dy, \qquad \text{or} \qquad a_n = \begin{cases} 0, & \text{if } n \text{ is even}, \\ \frac{4V}{n\pi}, & \text{if } n \text{ is odd}. \end{cases}$$

Therefore, the solution has the form
$$u(x,y) = \frac{4V}{\pi} \sum_{k=0}^{\infty} \frac{\sinh\left[\frac{(2k+1)(a-x)\pi}{b}\right] \sin\left[\frac{(2k+1)\pi y}{b}\right]}{(2k+1)\sinh\left[\frac{(2k+1)\pi a}{b}\right]}.$$

Example 2 [18]. Suppose that two sides, AC and BC, of a rectangular homogeneous plate (see Figure 1.2) are covered with a heat insulation, and the other two sides are maintained at temperature zero. Find the stationary temeperature distribution in the plate under the assumption that a quantity of heat $Q = \text{const}$ is extracted it.

Solution. We are dealing with a boundary value problem for the Poisson equation with boundary conditions of mixed type;

$$
\begin{cases}
u_{xx} + u_{yy} = -\dfrac{Q}{k}, & 0 < x < a, \quad 0 < y < b, \\
u(0, y) = 0, \quad u_x(a, y) = 0, & 0 \le y \le b, \\
u(x, 0) = 0, \quad u_y(x, b) = 0, & 0 \le x \le a
\end{cases}
\tag{1.17}
$$

(here k is the internal heat conduction coefficient).

The eigenvalues and eigenfunctions of the problem are found by solving the auxiliary boundary value problem (Sturm-Liouville problem)

$$
\begin{cases}
X'' + \lambda^2 X = 0, & 0 < x < a, \\
X(0) = 0 = X'(a) = 0.
\end{cases}
$$

We get $\lambda_n^2 = \left[\dfrac{(2n+1)\pi}{2a} \right]^2$ and $X_n(x) = \sin\left[\dfrac{(2n+1)\pi}{2a} x \right]$, $n = 0, 1, \dots$ We will seek the solution of the above problem in the form of an expansion in eigenfunctions

$$
u(x, y) = \sum_{n=0}^{\infty} Y_n(y) \sin\left[\frac{(2n+1)\pi}{2a} x \right],
$$

where the functions $Y_n(y)$ are subject to determination. Substituting this expression of the solution in equation (1.17) we obtain

$$
-\sum_{n=0}^{\infty} Y_n(y) \frac{(2n+1)^2 \pi^2}{4a^2} \sin\left[\frac{(2n+1)\pi}{2a} x \right] + \sum_{n=0}^{\infty} Y_n''(y) \sin\left[\frac{(2n+1)\pi}{2a} x \right] =
$$

$$
= \sum_{n=0}^{\infty} \alpha_n \sin\left[\frac{(2n+1)\pi}{2a} x \right],
$$

where the Fourier coefficients α_n of the function $-Q/k$ are equal to

$$
\alpha_n = \frac{2}{a} \int_0^a \left(-\frac{Q}{k} \right) \sin\left[\frac{(2n+1)\pi}{2a} x \right] dx = -\frac{4Q}{k\pi(2n+1)} .
$$

This yields the following boundary value problem for the determination of the function $Y_n(y)$, $n = 0, 1, 2, \dots$:

$$
\begin{cases}
Y_n'' - \dfrac{(2n+1)^2 \pi^2}{4a^2} Y_n(y) = -\dfrac{4Q}{k\pi(2n+1)}, & 0 < y < b, \\
Y_n(0) = 0, \quad Y_n'(b) = 0.
\end{cases}
$$

Solving this problem, we obtain

$$Y_n(y) = a_n \cosh\left[\frac{(2n+1)\pi}{2a}y\right] + b_n \sinh\left[\frac{(2n+1)\pi}{2a}y\right] + \frac{16Qa^2}{k\pi^3(2n+1)^3},$$

where

$$a_n = -\frac{16Qa^2}{k\pi^3(2n+1)^3},$$

and

$$b_n = \frac{16Qa^2}{k\pi^3(2n+1)^3}\tanh\left[\frac{(2n+1)\pi b}{2a}y\right].$$

The final expression of the solution is

$$u(x,y) = \frac{16Qa^2}{k\pi^3}\sum_{n=0}^{\infty}\frac{1}{(2n+1)^3}\left(1 - \frac{\cosh\left[\frac{(2n+1)(b-y)\pi}{2a}\right]}{\cosh\left[\frac{(2n+1)\pi b}{2a}\right]}\right)\sin\left[\frac{(2n+1)\pi}{2a}x\right].$$

Example 3 [18]. Find the solution of the Laplace equation in the strip $0 \le x \le a$, $0 \le y < \infty$ which satisfies the boundary conditions

$$u(x,0) = 0, \quad u(a,y) = 0, \quad u(x,0) = A\left(1 - \frac{x}{a}\right), \quad u(x,\infty) = 0.$$

Solution. Thus, we need to solve the boundary value problem

$$\begin{cases} u_{xx} + u_{yy} = 0, & 0 < x < a, \quad 0 < y < \infty, \\ u(0,y) = u(a,y) = 0, & 0 \le y < \infty, \\ u(x,0) = A\left(1 - \frac{x}{a}\right), & u(x,\infty) = 0, \quad 0 \le x \le a \end{cases} \qquad (1.18)$$

Let us begin by finding the solution of the auxiliary problem

$$\begin{cases} v_{xx} + v_{yy} = 0, & 0 < x < a, \quad 0 < y < \infty, \\ v(0,y) = v(a,y) = 0, & 0 \le y < \infty, \end{cases}$$

in the form $v(x,y) = X(x)Y(y)$. We obtain two ordinary differential equations: (1) $X'' + \lambda X = 0$, and (2) $Y'' - \lambda Y = 0$.

From the conditions $v(0,y) = 0$, $v(a,y) = 0$ it follows that $X(0) = X(a) = 0$. Hence, the Sturm-Liouville problem

$$\begin{cases} X'' + \lambda X = 0, & 0 < x < a, \\ X(0) = X(a) = 0 \end{cases}$$

yields $\lambda_n = \left(\frac{n\pi}{a}\right)^2$ and $X_n(x) = \sin\left(\frac{n\pi}{a} x\right)$, $n = 1, 2, \ldots$. Then the corresponding solutions of the equation $Y'' - \lambda Y = 0$ are

$$Y_n(y) = A_n e^{-\frac{n\pi}{a} y} + B_n e^{\frac{n\pi}{a} y}.$$

We conclude that

$$v_n(x, y) = \left[A_n e^{-\frac{n\pi}{a} y} + B_n e^{\frac{n\pi}{a} y}\right] \sin\left(\frac{n\pi}{a} x\right).$$

Therefore, the solution of problem (1.18) is given by a series

$$u(x, y) = \sum_{n=1}^{\infty} \left[A_n e^{-\frac{n\pi}{a} y} + B_n e^{\frac{n\pi}{a} y}\right] \sin\left(\frac{n\pi}{a} x\right). \tag{1.19}$$

From the condition $u(x, \infty)$ it follows that $B_n = 0$, $n = 1, 2, \ldots$. Setting $y = 0$ in (1.19) we get

$$A\left(1 - \frac{x}{a}\right) = \sum_{n=1}^{\infty} A_n \sin\left(\frac{n\pi}{a} x\right),$$

i.e.,

$$A_n = \frac{2}{a} \int_0^a A\left(1 - \frac{x}{a}\right) \sin\left(\frac{n\pi}{a} x\right) dx = \frac{2A}{\pi n}.$$

We conclude that

$$u(x, y) = \frac{2A}{\pi} \sum_{n=1}^{\infty} \frac{1}{n} e^{-\frac{n\pi}{a} y} \sin\left(\frac{n\pi}{a} x\right).$$

Remark 1. The boundary value problem for the Laplace (Poisson) equation in a rectangular parallelepiped is solved in a similar manner.

Remark 2. Let us assume that the mathematical model of a given physical phenomenon is such that both the equation itself and the boundary conditions are inhomogeneous. Then by using the superposition principle the original boundary value problem can be decomposed into subproblems; one then solves the subproblems and adds their solutions to obtain the solution of the original problem.

For example, the solution of the Dirichlet problem

$$\begin{cases} \Delta u = f & \text{in the domain } \Omega, \\ u = \varphi & \text{on the boundary } \partial\Omega, \end{cases}$$

is the sum of the solutions of the following simpler problems:

(1) $\begin{cases} \Delta u = f & \text{in the domain } \Omega, \\ u = 0 & \text{on the boundary } \partial\Omega, \end{cases}$ (2) $\begin{cases} \Delta u = 0 & \text{in the domain } \Omega, \\ u = \varphi & \text{on the boundary } \partial\Omega. \end{cases}$

1.8. Boundary value problems for the Laplace and Poisson equations in a bounded cylinder

To treat the problems mentioned in the title we must resort to special functions, more precisely, to Bessel functions.

First let us consider a boundary value problem for the Laplace equation in a cylinder.

Example 1 [4, Ch. IV, no. 110]. Find the potential of the electrostatic field of a cylindrical wire of section $\rho \leq a$, $0 \leq z \leq l$, such that both bases of the cylinder are grounded and its lateral surface is charged at a potential V_0. Calculate the field intensity on the axis (Figure 1.3).

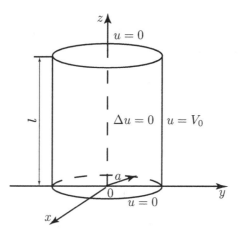

FIGURE 1.3.

Solution. We need to solve the Laplace equation inside the cylinder with given boundary conditions:

$$\begin{cases} \dfrac{1}{\rho}\dfrac{\partial}{\partial\rho}\left(\rho\dfrac{\partial u}{\partial\rho}\right) + \dfrac{\partial^2 u}{\partial z^2} = 0, & 0 < \rho < a, \quad 0 < z < l, \\ u(\rho,0) = u(\rho,l) = 0, & 0 \leq \rho \leq a, \\ u(a,z) = V_0, & 0 \leq z \leq l \end{cases}$$

(the solution $u(\rho,z)$ does not depend on φ since the boundary values are independent of φ). Using the method of separation of variables, we represent the solution in the form $u(\rho,z) = R(\rho)Z(z)$. Substituting this expression in the Laplace equation

$$\frac{1}{\rho}\frac{\partial}{\partial\rho}\left(\rho\frac{\partial u}{\partial\rho}\right) + \frac{\partial^2 u}{\partial z^2} = 0$$

we get

$$Z\rho\frac{\partial}{\partial\rho}(\rho R') + RZ'' = 0$$

whence, upon dividing both sides by RZ,

$$\frac{\rho\frac{\partial}{\partial\rho}(\rho R')}{R} + \frac{Z''}{Z} = 0,$$

or

$$\frac{\rho\frac{\partial}{\partial\rho}(\rho R')}{R} = -\frac{Z''}{Z} = \lambda, \tag{1.20}$$

where λ is the separation constant. Clearly, on physical grounds $\lambda > 0$: otherwise the function $Z(z)$, and together with it the potential, would not vanish on the upper and bottom bases of the cylindrical wire.

Equation (1.20) yields two ordinary differential equations:

$$(1)\quad Z'' + \lambda Z = 0,$$

and

$$(2)\quad \frac{1}{\rho}\frac{d}{d\rho}(\rho R') - \lambda R = 0.$$

Using the fact that $Z(0) = Z(l) = 0$, we obtain the standard Sturm-Liouville problem:

$$\begin{cases} Z'' + \lambda Z = 0, & 0 < z < l, \\ Z(0) = Z(l) = 0. \end{cases}$$

This problem has the eigenfunctions $Z_n(z) = \sin\left(\frac{n\pi}{l}z\right)$, corresponding to the eigenvalues $\lambda_n = \left(\frac{n\pi}{l}\right)^2$, $n = 1, 2, \ldots$. The function $R(\rho)$ is determinded from the equation

$$\frac{1}{\rho}\frac{d}{d\rho}(\rho R') - \left(\frac{n\pi}{l}\right)^2 R = 0, \tag{1.21}$$

which is recognized to be the Bessel equation of index zero and imaginary argument. Indeed, from equation (1.21) it follows that

$$\rho^2 R'' + \rho R' - \rho^2 \left(\frac{n\pi}{l}\right)^2 R = 0.$$

Passing in this equation to the new independent variable $x = \rho\frac{n\pi}{l}$ and using the relations

$$R' = \frac{dR}{dx}\frac{n\pi}{l}, \quad R'' = \frac{d^2R}{dx^2}\left(\frac{n\pi}{l}\right)^2,$$

we arrive at the equation

$$x^2 \frac{d^2 R}{dx^2} + x \frac{dR}{dx} - x^2 R = 0.$$

Its general solution is written in the form

$$R(x) = C_1 I_0(x) + C_2 K_0(x),$$

where $I_0(x)$ and $K_0(x)$ are the Bessel functions of index zero and imaginary argument, of the first and second kind, respectively, and C_1 and C_2 are arbitrary constants. Since (the Macdonald) function $K_0(x) \to \infty$ when $x \to 0$, we must set $C_2 = 0$ (otherwise the solution of our problem will be unbounded on the axis of the cylinder). Therefore,

$$R_n(\rho) = C I_0 \left(\frac{n\pi}{l} \rho \right).$$

The "atoms" from which the solution of the original problem will be constructed are the functions

$$I_0 \left(\frac{n\pi}{l} \rho \right) \sin \left(\frac{n\pi}{l} z \right), \qquad n = 1, 2, \ldots$$

Thus, the solution of our has the series representation

$$u(\rho, z) = \sum_{n=1}^{\infty} c_n I_0 \left(\frac{n\pi}{l} \rho \right) \sin \left(\frac{n\pi}{l} z \right).$$

The constants c_n are found from the boundary condition $u(a, z) = V_0$. We have

$$V_0 = \sum_{n=1}^{\infty} c_n I_0 \left(\frac{n\pi}{l} a \right) \sin \left(\frac{n\pi}{l} z \right),$$

whence

$$c_n I_0 \left(\frac{n\pi}{l} a \right) = \frac{2}{l} \int_0^l V_0 \sin \left(\frac{n\pi}{l} z \right) dz = \begin{cases} \dfrac{4V_0}{n\pi}, & n \text{ is odd}, \\ 0, & n \text{ is even}. \end{cases}$$

We conclude that

$$u(z, \rho) = \frac{4V_0}{\pi} \sum_{k=0}^{\infty} \frac{I_0 \left[\dfrac{(2k+1)\pi}{l} \rho \right]}{I_0 \left[\dfrac{(2k+1)\pi}{l} a \right]} \cdot \frac{\sin \left[\dfrac{(2k+1)\pi}{l} z \right]}{2k+1}.$$

The field on the axis of the cylinder is

$$E_z(0, z) = -\frac{\partial u}{\partial z}(0, z) = -\frac{4V_0}{l} \sum_{k=0}^{\infty} \frac{\cos \left[\dfrac{(2k+1)\pi}{l} z \right]}{I_0 \left[\dfrac{(2k+1)\pi}{l} a \right]}.$$

Example 2 [18]. Consider a cylinder with base of radius R and height h. Assume that the temperature of the lower base and of the lateral surface is equal to zero, while the temperature of the upper base is a given function of ρ. Find the steady temperature distribution in the interior of the cylinder.

Solution. The mathematical formulation of the problems is as follows:

$$\begin{cases} \dfrac{1}{\rho}\dfrac{\partial}{\partial\rho}\left(\rho\dfrac{\partial u}{\partial\rho}\right) + \dfrac{\partial^2 u}{\partial z^2} = 0, & 0 < \rho < R, \quad 0 < z < h, \\ u(\rho,0) = 0, \quad u(\rho,h) = f(\rho), & 0 \le \rho \le R, \\ u(R,z) = 0, & 0 \le z \le h. \end{cases}$$

Setting, as before, $u(\rho,z) = r(\rho)Z(z)$ and substituting this expression in the Laplace equation, we obtain two ordinary differential equations:

$$\begin{aligned} &(1) \quad \frac{1}{\rho}\frac{d}{d\rho}(\rho r') + \lambda r = 0; \\ &(2) \quad Z'' - \lambda Z = 0. \end{aligned} \qquad (1.22)$$

We note that here $\lambda > 0$ (this will be clear once we find the solution). The boundary condition $u(R,z) = 0$ implies $r(R) = 0$. Equation (1.22) can be rewritten as

$$\rho^2 r'' + \rho r' + \lambda \rho^2 r = 0. \qquad (1.23)$$

Passing to the new independent variable $x = \sqrt{\lambda}\rho$ we obtain the Bessel equation of order zero

$$x^2 \frac{d^2 r}{dx^2} + x\frac{dr}{dx} + x^2 r = 0,$$

whose general solution has the form

$$r(x) = C_1 J_0(x) + C_2 B_0(x),$$

where $J_0(x)$ and $B_0(x)$ are the Bessel function of order zero of first and second kind, respectively, and C_1, C_2 are arbitrary constants.

Returning to the old variable ρ we have

$$r(\rho) = C_1 J_0(\sqrt{\lambda}\rho) + C_2 B_0(\sqrt{\lambda}\rho).$$

Thus, in the present case solving the Sturm-Liouville problem

$$\begin{cases} \rho^2 r'' + \rho r' + \lambda \rho^2 r = 0, & 0 < \rho < R, \\ |r(0)| < \infty, \quad r(R) = 0 \end{cases}$$

reduces to the solution of the Bessel equation with the indicated boundary conditions. Since $B_0(\sqrt{\lambda}\rho) \to \infty$ as $\rho \to 0$, we must set $C_2 = 0$, and so $r(\rho) =$

$CJ_0(\sqrt{\lambda}\rho)$. From the condition $r(R) = 0$ it follows that $J_0(\sqrt{\lambda}R) = 0$. Denoting by $\mu_1, \mu_2, \ldots, \mu_n, \ldots$ the positive roots of the Bessel function $J_0(x)$ (Figure 1.4), we obtain the eigenvalues $\lambda_n = \left(\frac{\mu_n}{R}\right)^2$ and the corresponding eigenfunctions $J_0\left(\frac{\mu_n}{R}\rho\right)$, $n = 1, 2, \ldots$. Further, from the equation (2) in (1.22) with $\lambda = \lambda_n = \left(\frac{\mu_n}{R}\right)^2$ we obtain

$$Z_n(x) = A_n\cosh\left(\frac{\mu_n}{R}z\right) + B_n\sinh\left(\frac{\mu_n}{R}z\right),$$

where A_n and B_n are arbitrary constants. From the boundary condition $u(\rho, 0) = 0$ it follows that $Z(0) = 0$, i.e., $A_n = 0$ for all n. Therefore, the "atoms" of the sought solution are the functions

$$J_0\left(\frac{\mu_n}{R}\rho\right)\sinh\left(\frac{\mu_n}{R}z\right), \qquad n = 1, 2\ldots.$$

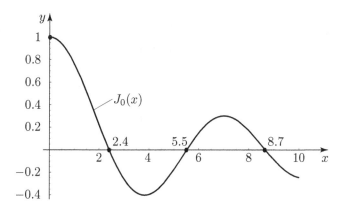

FIGURE 1.4.

The solution of our problem is given by a series

$$u(\rho, z) = \sum_{n=1}^{\infty} B_n J_0\left(\frac{\mu_n}{R}\rho\right)\sinh\left(\frac{\mu_n}{R}z\right).$$

The constants B_n are found from the boundary condition $u(\rho, h) = f(\rho)$. Indeed, we have

$$u(\rho, h) = \sum_{n=1}^{\infty} B_n J_0\left(\frac{\mu_n}{R}\rho\right)\sinh\left(\frac{\mu_n}{R}h\right),$$

or

$$f(\rho) = \sum_{n=1}^{\infty} B_n J_0\left(\frac{\mu_n}{R}\rho\right)\sinh\left(\frac{\mu_n}{R}h\right).$$

Multiplying both sides of this equality by $\rho J_0 \left(\frac{\mu_m}{R} \rho \right)$ and integrating the result over the segment $[0, R]$ we get

$$\int_0^R \rho f(\rho) J_0 \left(\frac{\mu_m}{R} \rho \right) d\rho = B_m \sinh \left(\frac{\mu_m}{R} h \right) \int_0^R \rho J_0^2 \left(\frac{\mu_m}{R} \rho \right) d\rho.$$

But

$$\int_0^R \rho J_0^2 \left(\frac{\mu_m}{R} \rho \right) d\rho = \frac{R^2}{2} J_1^2(\mu_m),$$

where $J_1(x)$ is the Bessel function of first kind and order one. Therefore, the solution of the problem has the form

$$u(\rho, z) = \frac{2}{R^2} \sum_{n=1}^{\infty} \frac{\sinh \left(\frac{\mu_n}{R} z \right)}{\sinh \left(\frac{\mu_n}{R} h \right)} \frac{J_0 \left(\frac{\mu_n}{R} \rho \right)}{J_1^2(\mu_n)} \int_0^R \rho f(\rho) J_0 \left(\frac{\mu_n}{R} \rho \right) d\rho.$$

Example 3. Find the potential in the interior points of a grounded cylinder of height h and with base of radius R, given that in the cylinder there is a charge distribution with density $\gamma = A z J_0 \left(\frac{\mu_3}{R} \rho \right)$ (where A is a constant).

Solution. We must solve the Poisson equation with null boundary conditions:

$$\begin{cases} \dfrac{1}{\rho} \dfrac{\partial}{\partial \rho} \left(\rho \dfrac{\partial u}{\partial \rho} \right) + \dfrac{\partial^2 u}{\partial z^2} = -4\pi A z J_0 \left(\dfrac{\mu_3}{R} \rho \right), \\ 0 < \rho < R, \quad 0 < z < h, \\ u(\rho, 0) = u(\rho, h) = 0, \qquad 0 \le \rho \le R, \\ u(R, z) = 0, \qquad 0 \le z \le h. \end{cases} \tag{1.24}$$

Let us seek the solution in the form $u(\rho, z) = J_0 \left(\frac{\mu_3}{R} \rho \right) f(z)$, where the function $f(z)$ is subject to determination. Substituting this expression of $u(\rho, z)$ in equation (1.24) we get

$$\frac{1}{\rho} \frac{d}{d\rho} \left[\rho \frac{d}{d\rho} J_0 \left(\frac{\mu_3}{R} \rho \right) \right] f(z) + J_0 \left(\frac{\mu_3}{R} \rho \right) f''(z) = -4\pi A z J_0 \left(\frac{\mu_3}{R} \rho \right). \tag{1.25}$$

Now let us observe that the function $J_0 \left(\frac{\mu_3}{R} \rho \right)$ is an eigenfunction of the Bessel equation, i.e.,

$$\frac{1}{\rho} \frac{d}{d\rho} \left[\rho \frac{d}{d\rho} J_0 \left(\frac{\mu_3}{R} \rho \right) \right] + \frac{\mu_3^2}{R^2} J_0 \left(\frac{\mu_3}{R} \rho \right) = 0.$$

Consequently, (1.25) gives

$$-\left(\frac{\mu_3}{R} \right)^2 J_0 \left(\frac{\mu_3}{R} \rho \right) f(z) + J_0 \left(\frac{\mu_3}{R} \rho \right) f''(z) = -4\pi A z J_0 \left(\frac{\mu_3}{R} \rho \right),$$

which in turn yields the following ordinary differential equation for the determination of $f(z)$:

$$f'' - \left(\frac{\mu_3}{R}\right)^3 f = -4\pi Az, \qquad 0 < z < h,$$

with $f(0) = f(h) = 0$. Solving this boundary value problem we find that

$$f(z) = -\frac{4\pi A R^2 h}{\mu_3^2} \cdot \frac{\sinh\left(\frac{\mu_3}{R} z\right)}{\sinh\left(\frac{\mu_3}{R} h\right)} + \frac{4\pi A R^2}{\mu_3^2} z.$$

Thus, the solution of our problem is given by the expression

$$u(\rho, z) = J_0\left(\frac{\mu_3}{R} \rho\right) \frac{4\pi A R^2}{\mu_3^2} \left[h \frac{\sinh\left(\frac{\mu_3}{R} z\right)}{\sinh\left(\frac{\mu_3}{R} h\right)} - z\right].$$

1.9. Boundary value problems for the Laplace and Poisson equations in a ball

To deal with the problem mentioned in the title we need to use spherical functions and solid spherical harmonics.

Recall that the general solution of the Laplace equation has the following form (in spherical coordinates (ρ, θ, φ)):

(1) $u(\rho, \theta, \varphi) = \sum_{n=0}^{\infty} \left(\frac{\rho}{a}\right)^n Y_n(\theta, \varphi)$ in the interior the sphere of radius a;

(2) $u(\rho, \theta, \varphi) = \sum_{n=0}^{\infty} \left(\frac{\rho}{a}\right)^{(n+1)} Y_n(\theta, \varphi)$ in the exterior of the sphere of radius a;

(3) $u(\rho, \theta, \varphi) = \sum_{n=0}^{\infty} \left(A_n \rho^n + \frac{B_n}{\rho^{n+1}}\right) Y_n(\theta, \varphi)$ in a spherical layer.

Here

$$Y_n(\theta, \varphi) = \sum_{m=0}^{n} [A_{nm} \cos(m\varphi) + B_{nm} \sin(m\varphi)] P_n^{(m)}(\cos \theta),$$

where $P_n^{(m)}(x)$ are the so-called associated Legendre functions.

Example 1. Find the solution $u(\rho, \theta, \varphi)$ of the interior Dirichlet problem for the Laplac equation with the boundary condition $u(a, \theta, \varphi) = \sin(3\theta) \cos \varphi$.

Solution. In spherical coordinates the problem is written as follows:

$$\begin{cases} \frac{1}{\rho^2} \frac{\partial}{\partial \rho}\left(\rho^2 \frac{\partial u}{\partial \rho}\right) + \frac{1}{\rho^2 \sin \theta} \frac{\partial}{\partial \theta}\left(\sin \theta \frac{\partial u}{\partial \theta}\right) + \frac{1}{\rho^2 \sin^2 \theta} \frac{\partial^2 u}{\partial \varphi^2} = 0, \\ 0 < \rho < a, \quad 0 < \theta < \pi, \quad 0 \le \varphi < 2\pi, \\ u(a, \theta, \varphi) = \sin(3\theta) \cos \varphi, \qquad 0 \le \theta \le \pi, \quad 0 \le \varphi \le 2\pi. \end{cases} \qquad (1.26)$$

Setting $u(\rho, \theta, \varphi) = R(\rho)Y(\theta, \varphi)$ and substituting this expression in equation (1.26), we obtain

$$Y\frac{d}{d\rho}(\rho^2 R') + R\left[\frac{1}{\sin\theta}\frac{\partial}{\partial\theta}\left(\sin\theta\frac{\partial Y}{\partial\theta}\right) + \frac{1}{\sin^2\theta}\frac{\partial^2 Y}{\partial\varphi^2}\right] = 0,$$

which upon dividing both sides by RY yields

$$\frac{\frac{d}{d\rho}(\rho^2 R')}{R} + \frac{\frac{1}{\sin\theta}\frac{\partial}{\partial\theta}\left(\sin\theta\frac{\partial Y}{\partial\theta}\right) + \frac{1}{\sin^2\theta}\frac{\partial^2 Y}{\partial\varphi^2}}{Y} = 0,$$

or

$$\frac{\frac{d}{d\rho}(\rho^2 R')}{R} = -\frac{\frac{1}{\sin\theta}\frac{\partial}{\partial\theta}\left(\sin\theta\frac{\partial Y}{\partial\theta}\right) + \frac{1}{\sin^2\theta}\frac{\partial^2 Y}{\partial\varphi^2}}{Y} = \lambda,$$

where λ is the separation constant. This yields two equations:

$$
\begin{aligned}
&(1)\quad \rho^2 R'' + 2\rho R' - \lambda R = 0,\\
&(2)\quad \frac{1}{\sin\theta}\frac{\partial}{\partial\theta}\left(\sin\theta\frac{\partial Y}{\partial\theta}\right) + \frac{1}{\sin^2\theta}\frac{\partial^2 Y}{\partial\varphi^2} + \lambda Y = 0;
\end{aligned}
\tag{1.27}
$$

here the function $Y(\theta, \varphi)$ must be restricted to the sphere.

Moreover, the function $Y(\theta, \varphi)$ satisfies the conditions

$$
\begin{cases}
Y(\theta, \varphi) = Y(\theta, \varphi + 2\pi),\\
|Y(0, \varphi)| < \infty, \quad |Y(\pi, \varphi)| < \infty.
\end{cases}
\tag{1.28}
$$

As is known, the bounded solutions of equation (1.27) that have continuous derivatives up to and including order two are called *spherical functions*.

The solution of problem (1.27),(1.28) for $Y(\theta, \varphi)$ will also be sought via separation of variables, setting $Y(\theta, \varphi) = T(\theta)\Phi(\varphi)$. Susbtituting this expression in equation (1.27), we get

$$\Phi\frac{1}{\sin\theta}\frac{d}{d\theta}(\sin\theta T') + \frac{1}{\sin^2\theta}T\Phi'' + \lambda T\Phi = 0,$$

whence

$$\frac{\sin\theta\frac{d}{d\theta}(\sin\theta T')}{T} + \lambda\sin^2\theta = -\frac{\Phi''}{\Phi} = \mu.$$

Thus, the function $\Phi(\varphi)$ is found by solving the problem

$$
\begin{cases}
\Phi'' + \mu\Phi = 0,\\
\Phi(\varphi) = \Phi(\varphi + 2\pi).
\end{cases}
$$

We have already solved such a problem when we considered the Laplace equation in a disc, and found that $\mu = m^2$ and $\Phi_m(\varphi) = C_1 \cos(m\varphi) + C_2 \sin(m\varphi)$, where C_1 and C_2 are arbitrary constants and $m = 0, 1, \ldots$

The function $T(\theta)$ is found from the equation

$$\frac{1}{\sin\theta} \frac{d}{d\theta}(\sin\theta T') + \left(\lambda - \frac{m^2}{\sin^2\theta}\right) T = 0 \qquad (1.29)$$

and the conditions that T be bounded at $\theta = 0$ and $\theta = \pi$. Introducing the new variable $x = \cos\theta$ and observing that

$$T' = \frac{dT}{dx} \frac{dx}{d\theta} = \frac{dT}{dx}(-\sin\theta),$$

$$T'' = \frac{d^2T}{dx^2} \sin^2\theta - \frac{dT}{dx} \cos\theta,$$

equation (1.29) yields the following boundary value problem for eigenvalues and eigenfunctions:

$$\begin{cases} (1 - x^2)\dfrac{d^2T}{dx^2} - 2x\dfrac{dT}{dx} + \left(\lambda - \dfrac{m^2}{1 - x^2}\right) T = 0, \quad -1 < x < 1, \\ |T(-1)| < \infty, \quad |T(+1)| < \infty. \end{cases}$$

The eigenfunctions of this problem,

$$T_n^{(m)}(x) = P_n^{(m)}(x) = (1 - x^2)^{m/2} \frac{d^m}{dx^m} P_n(x),$$

are the associated Legendre functions. Hence, the solutions of equation (1.29) are the functions $T_n^{(m)}(x) = P_n^{(m)}(\cos\theta)$.

Combining the solutions of equation (1.29) with the solutions of the equation $\Phi'' + \mu\Phi = 0$, we obtain the $2n + 1$ spherical functions

$$P_n(\cos\theta), \quad P_n^{(m)}(\cos\theta)\cos(m\varphi), \quad P_n^{(m)}(\cos\theta)\sin(m\varphi),$$
$$n = 0, 1, \ldots; \quad m = 1, 2, \ldots$$

The general solution of equation (1.27) for $\lambda = n(n+1)$ is written in the form

$$Y_n(\theta, \varphi) = \sum_{m=0}^{n} [A_{nm} \cos(m\varphi) + B_{nm} \sin(m\varphi)] P_n^{(m)}(\cos\theta).$$

Now let us return to the search for the function $R(\rho)$. Setting $R(\rho) = \rho^\sigma$ and substituting this expression in the equation $\rho^2 R'' + 2\rho R' - \lambda R = 0$, we obtain

$\sigma(\sigma+1) - n(n+1) = 0$, whence $\sigma_1 = n$, $\sigma_2 = -(n+1)$. Thus, the solution "atoms" are the functions

$$\rho^n P_n^{(m)}(\cos\theta)\cos(m\varphi), \qquad \rho^n P_n^{(m)}(\cos\theta)\sin(m\varphi),$$
$$\rho^{-(n+1)} P_n^{(m)}(\cos\theta)\cos(m\varphi), \qquad \rho^{-(n+1)} P_n^{(m)}(\cos\theta)\sin(m\varphi).$$

However, the solutions $\rho^{-(n+1)} P_n^{(m)}(\cos\theta)\cos(m\varphi)$, $\rho^{-(n+1)} P_n^{(m)}(\cos\theta)\sin(m\varphi)$ must be discarded because they are not bounded when $\rho \to 0$. Hence, the solution of our problem is given by a series

$$u(\rho,\theta,\varphi) = \sum_{n=0}^{\infty}\sum_{m=0}^{n} \rho^n \left[A_{nm}\cos(m\varphi) + B_{nm}\sin m\varphi\right] P_n^{(m)}(\cos\theta).$$

It remains to choose the constants A_{nm} and B_{nm} so that the boundary condition

$$u(a,\theta,\varphi) = \sin(3\theta)\cos\varphi$$

will be satisfied. We have

$$u(a,\theta,\varphi) = \sum_{n=0}^{\infty}\sum_{m=0}^{n} a^n \left[A_{nm}\cos(m\varphi) + B_{nm}\sin m\varphi\right] P_n^{(m)}(\cos\theta),$$

i.e., we must satisfy the equality

$$\sin(3\theta)\cos\varphi = \sum_{n=0}^{\infty}\sum_{m=0}^{n} a^n \left[A_{nm}\cos(m\varphi) + B_{nm}\sin m\varphi\right] P_n^{(m)}(\cos\theta).$$

It follows that in the sum $\sum_{m=0}^{n}\cdots$ we must retain only the term corresponding to $m = 1$. This yields

$$\sin(3\theta) = \sum_{n=1}^{\infty} a^n A_{n1} P_n^{(1)}(\cos\theta).$$

The coefficients A_{n1} can be found from the general formula: if

$$f(\theta) = \sum_{n=1}^{\infty} b_n P_n^{(1)}(\cos\theta),$$

then

$$b_n = \frac{2n+1}{2} \cdot \frac{(n-1)!}{(n+1)!} \int_0^\pi f(\theta) P_n^{(1)}(\cos\theta)\sin\theta\, d\theta.$$

However, it is more convenient to proceed as follows: we have

$$\sin(3\theta) = \sin\theta(4\cos^2\theta - 1), \quad P_n^{(1)}(\cos\theta) = \sin\theta\frac{dP_n(\cos\theta)}{d(\cos\theta)},$$

$$P_1(x) = x, \quad P_3(x) = \frac{1}{2}(5x^3 - 3x).$$

Therefore,

$$(4\cos^2\theta - 1)\sin\theta = \sin\theta\left[a\cdot A_{11}\cdot 1 + a^3\cdot A_{31}\cdot\frac{1}{2}(15\cos^2\theta - 3)\right],$$

which gives

$$A_{11} = -\frac{1}{5a}, \quad A_{31} = \frac{8}{15a^3}, \quad A_{n1} = 0, \quad n = 2, 4, 5, \ldots$$

We conclude that the solution of our problem has the form

$$u(\rho, \theta, \varphi) = \left(-\frac{1}{5}\right)\frac{\rho}{a}P_1^{(1)}(\cos\theta)\cos\varphi + \frac{8}{15}\left(\frac{\rho}{a}\right)^3 P_3^{(1)}(\cos\theta)\cos\varphi.$$

Example 2. Find a function u, harmonic inside the spherical layer $R_1 < \rho < R_2$, and such that

$$u|_{\rho=R_1} = P_2^{(1)}(\cos\theta)\sin\varphi, \quad u|_{\rho=R_2} = P_5^{(3)}(\cos\theta)\cos(3\varphi).$$

Solution. The mathematical formulation of the problem is

$$\begin{cases} \Delta u = 0, \quad R_1 < \rho < R_2, \quad 0 < \theta < \pi, \quad 0 < \varphi < 2\pi, \\ u(R_1, \theta, \varphi) = P_2^{(1)}(\cos\theta)\sin\varphi, \\ u(R_2, \theta, \varphi) = P_5^{(3)}(\cos\theta)\cos(3\varphi), \end{cases} \quad 0 \le \theta \le \pi, \quad 0 \le \varphi \le 2\pi$$

(see Figure 1.5).

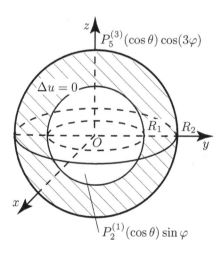

FIGURE 1.5.

The solution of this problem is written in the form

$$u(\rho,\theta,\varphi) = \sum_{n=0}^{\infty}\sum_{m=0}^{n}\left[\left(A_{nm}\rho^n + \frac{B_{nm}}{\rho^{n+1}}\right)\cos(m\varphi)+\right.$$

$$\left.+ \left(C_{nm}\rho^n + \frac{D_{nm}}{\rho^{n+1}}\right)\sin(m\varphi)\right]P_n^{(m)}(\cos\theta),$$

where the numbers A_{nm}, B_{nm}, C_{nm} and D_{nm} are subject to determination. The boundary conditions yield the following systems of equations for the coefficients of the expansion:

$$(1)\quad\begin{cases} C_{21}R_1^2 + \dfrac{D_{21}}{R_1^3} = 1, \\[2mm] A_{21}R_1^2 + \dfrac{B_{21}}{R_1^3} = 0, \\[2mm] C_{21}R_2^2 + \dfrac{D_{21}}{R_2^3} = 0, \\[2mm] A_{21}R_2^2 + \dfrac{B_{21}}{R_2^3} = 0, \end{cases}\qquad(2)\quad\begin{cases} A_{53}R_1^5 + \dfrac{B_{53}}{R_1^6} = 0, \\[2mm] C_{53}R_1^5 + \dfrac{D_{53}}{R_1^6} = 0, \\[2mm] A_{53}R_2^5 + \dfrac{B_{53}}{R_2^6} = 1, \\[2mm] C_{53}R_2^5 + \dfrac{D_{53}}{R_2^6} = 0, \end{cases}$$

All the remaining coefficients are equal to zero. Solving the above systems we obtain

$$A_{21} = B_{21} = 0, \qquad C_{53} = D_{53} = 0, \qquad C_{21} = -\frac{R_1^3}{R_2^2(R_2^5 - R_1^5)},$$

$$D_{21} = \frac{(R_1 R_2)^3}{R_2^5 - R_1^5}, \qquad A_{53} = -\frac{R_2^6}{R_1^5(R_2^{11} - R_1^{11})}, \qquad B_{53} = \frac{(R_1 R_2)^6}{R_2^{11} - R_1^{11}}.$$

Therefore, the harmonic function sought has the form

$$u(\rho,\theta,\varphi) = \left(C_{21}\rho + \frac{D_{21}}{\rho^2}\right)P_2^{(1)}(\cos\theta)\sin\varphi+$$

$$+ \left(A_{53}\rho^5 + \frac{B_{53}}{\rho^6}\right)P_5^{(3)}(\cos\theta)\cos(3\varphi).$$

Example 3 [6, 16.25(1)]. Find a function u, harmonic inside the spherical layer $1 < \rho < 2$, such that

$$\left(3u + \frac{\partial u}{\partial\rho}\right)\Bigg|_{\rho=1} = 5\sin^2\theta\sin(2\varphi) \qquad \text{and} \qquad u|_{\rho=2} = -\cos\theta.$$

Solution. The problem is formulated mathematically as follows:

$$
\begin{cases}
\Delta u = 0, \quad 1 < \rho < 2, \quad 0 < \theta < \pi, \quad 0 \leq \varphi < 2\pi, \\
\left(3u + \dfrac{\partial u}{\partial \rho} \right)\bigg|_{\rho=1} = 5\sin^2\theta\sin(2\varphi), \quad 0 \leq \theta \leq \pi, \quad 0 \leq \varphi \leq 2\pi, \\
u|_{\rho=2} = -\cos\theta, \quad 0 \leq \theta \leq \pi, \quad 0 \leq \varphi < 2\pi.
\end{cases}
$$

We have

$$
u(\rho, \theta, \varphi) = \sum_{n=0}^{\infty} \sum_{m=0}^{\infty} \left[\left(A_{nm}\rho^n + \frac{B_{nm}}{\rho^{n+1}} \right) \cos(m\varphi) + \right.
$$

$$
\left. + \left(C_{nm}\rho^n + \frac{D_{nm}}{\rho^{n+1}} \right) \sin(m\varphi) \right] P_n^{(m)}(\cos\theta).
$$

From the boundary conditions it follows that in this sum we must retain only the terms with the indices $n = 2$, $m = 2$ and $n = 1$, $m = 0$. In other words, it is convenient to seek the solution in the form

$$
u(\rho, \theta, \varphi) = \left(a\rho + \frac{b}{\rho^2} \right) \cos\theta + \left(c\rho^2 - \frac{d}{\rho^3} \right) \sin^2\theta\sin(2\varphi).
$$

Using the boundary conditions we obtain the following system of equations for the determination of the coefficients a, b, c, d:

$$
\begin{cases}
4a + b = 0, \\
5c = 5, \\
2a + b/4 == -1, \\
4c - d/8 = 0.
\end{cases}
$$

Solving this system, we obtain $a = -1$, $b = 4$, $c = 1$, $d = 32$. Hence, the solution has the expression

$$
u(\rho, \theta, \varphi) = \left(-\rho + \frac{4}{\rho^2} \right) \cos\theta + \left(\rho^2 - \frac{32}{\rho^3} \right) \sin^2\theta\sin(2\varphi).
$$

Example 4 [4, Ch. IV, no. 125]. Find the solution of the Neumann problem for the Laplace equation in the interior of the sphere of radius a with the condition

$$
\frac{\partial u}{\partial n}(a, \theta, \varphi) = A\cos\theta \quad (A = \text{const}).
$$

Solution. We are dealing with the case of an axially-symmetric solution of the Neumann problem for the Laplace equation, since the boundary condition does not depend on φ, and consequently the solution also does not depend of φ: $u = u(\rho, \theta)$.

First of all, it is readily verified that the necessary condition for the solvability of our problem is satisfied. Indeed

$$\int_0^{2\pi} \int_0^{\pi} \frac{\partial u}{\partial n}\, ds = 0, \qquad \text{or} \qquad \int_0^{2\pi} d\varphi \int_0^{\pi} A \cos\theta \sin\theta a^2\, d\theta = 0.$$

In the present case the Laplace equation has the form

$$\frac{\partial}{\partial \rho}\left(\rho^2 \frac{\partial u}{\partial \rho}\right) + \frac{1}{\sin\theta}\frac{\partial}{\partial\theta}\left(\sin\theta\frac{\partial u}{\partial\theta}\right) = 0, \qquad 0 \le \rho < a, \quad 0 \le \theta \le \pi.$$

Setting $u(\rho,\theta) = R(\rho)T(\theta)$ and substituting this expression in the equation, we obtain, after separation of variables, two ordinary differential equations:

$$\rho^2 R'' + 2\rho R' - \lambda R = 0, \tag{1.30}$$

and

$$\frac{1}{\sin\theta}\frac{d}{d\theta}(\sin\theta \cdot T') + \lambda T = 0. \tag{1.31}$$

If in the equation (1.31) we pass to the new variable $x = \cos\theta$ we arrive at the Legendre equation

$$\frac{d}{dx}\left[(1-x^2)\frac{dT}{dx}\right] + \lambda T = 0, \qquad -1 < x < 1, \tag{1.32}$$

under the condition $|T(\pm 1)| < \infty$. The bounded solutions of the Legendre equation (1.32) on the interval $(-1, 1)$ are the Legendre polynomials $P_n(x)$ for $\lambda_n = n(n+1)$. Hence, the bounded solutions of equation (1.31) on the interval $(0, \pi)$ are the functions $P_n(\cos\theta)$. The bounded solutions of equation (1.30) are the functions $R_n(\rho) = \rho^n$ $(n = 0, 1, 2, \dots)$. It follows that

$$u(\rho,\theta) = \sum_{n=0}^{\infty} C_n \rho^n P_n(\cos\theta),$$

where the constants C_n are to be determined from the boundary condition $\partial u/\partial\rho = A\cos\theta$. We have

$$\frac{\partial u}{\partial \rho}(\rho,\theta) = \sum_{n=0}^{\infty} nC_n a^{n-1} P_n(\cos\theta),$$

or, setting $\rho = a$,

$$A\cos\theta = \sum_{n=0}^{\infty} nC_n a^{n-1} P_n(\cos\theta)$$

whence, upon applying the formula

$$C_n = \frac{2n+1}{2na^{n-1}} \int_0^{\pi} A\cos\theta P_n(\cos\theta)\sin\theta\, d\theta,$$

we find that $C_1 = 1$ and $C_n = 0$ for $n = 2, 3, \ldots$. We conclude that

$$u(\rho, \theta) = C + A\rho \cos \theta,$$

where C is an arbitrary constant.

Example 3. Solve the following Dirichlet problem for the Poisson equation in a ball of radius a centered at the origin:

$$\begin{cases} \Delta u = xz & \text{in the interior of the ball,} \\ u|_{\rho=a} = 1. \end{cases}$$

Solution. Passing to spherical coordinates, we will seek the solution as a sum

$$u(\rho, \theta, \varphi) = v(\rho, \theta, \varphi) + w(\rho),$$

where the function $v(\rho, \theta, \varphi)$ is defined as the solution of the equation

$$\begin{cases} \dfrac{1}{\rho^2} \dfrac{\partial}{\partial \rho} \left(\rho^2 \dfrac{\partial v}{\partial \rho} \right) + \dfrac{1}{\rho^2} \dfrac{1}{\sin \theta} \dfrac{\partial}{\partial \theta} \left(\sin \theta \dfrac{\partial v}{\partial \theta} \right) + \\ \qquad + \dfrac{1}{\rho^2} \dfrac{1}{\sin^2 \theta} \dfrac{\partial^2 v}{\partial \varphi^2} = \dfrac{\rho^2}{2} \cos \varphi \sin(2\theta), \\ 0 < \rho < a, \quad 0 < \theta < a, \quad 0 \le \varphi < 2\pi, \\ v(a, \theta, \varphi) = 0, \end{cases} \tag{1.33}$$

and the function $w(\rho)$ is defined as the solution of the problem

$$\begin{cases} \dfrac{1}{\rho^2} \dfrac{d}{d\rho} (\rho^2 w') = 0, & 0 < \rho < a, \\ w(a) = 1, \quad |w(0)| < \infty. \end{cases} \tag{1.34}$$

Let us solve first problem (1.33), seeking the solution in the form

$$v(\rho, \theta, \varphi) = R(\rho) P_2^{(1)}(\cos \theta) \cos \varphi,$$

where $P_2^{(1)}(x)$ is the associated Legendre function with indices $n = 2$, $m = 1$. Substituting this expression of $v(\rho, \theta, \varphi)$ in the equation of problem (1.33) and denoting $P_2^{(1)}(\cos \theta) \cos \varphi = Y_2^{(1)}(\theta, \varphi)$ we get the equation

$$Y_2^{(1)} \frac{d}{d\rho}(\rho^2 R') + R \frac{1}{\sin \theta} \frac{\partial}{\partial \theta} \left(\sin \theta \frac{\partial Y_2^{(1)}}{\partial \theta} \right) + R \frac{1}{\sin^2 \theta} \frac{\partial^2 Y_2^{(1)}}{\partial \varphi^2} = \frac{\rho^4}{6} Y_2^{(1)}(\theta, \varphi).$$

But by the definition of the spherical function $Y_2^{(1)}(\theta, \varphi)$ one has the identity

$$\frac{1}{\sin \theta} \frac{\partial}{\partial \theta} \left(\sin \theta \frac{\partial Y_2^{(1)}}{\partial \theta} \right) + \frac{1}{\sin^2 \theta} \frac{\partial^2 Y_2^{(1)}}{\partial \varphi^2} + 6Y_2^{(1)} = 0, \quad 0 < \theta < \pi, \quad 0 < \varphi < 2\pi.$$

Therefore,

$$\frac{d}{d\rho}(\rho^2 R')Y_2^{(1)} - 6RY_2^{(1)} = \frac{\rho^4}{6}Y_2^{(1)},$$

which yields the equation

$$\frac{d}{d\rho}(\rho^2 R') - 6R = \frac{\rho^4}{6}, \qquad 0 < \rho < a,$$

together with the boundary conditions $|R(0)| < \infty$, $R(a) = 0$. Therefore, the function $R(\rho)$ is determined by solving the problem

$$\begin{cases} \rho^2 R'' + 2\rho R' - 6R = \dfrac{\rho^4}{6}, & 0 < \rho < a, \\ |R(0)| < \infty, \quad R(a) = 0. \end{cases}$$

Its solution is $R(\rho) = \frac{1}{84}\rho^2(\rho^2 - a^2)$. The solution of problem (1.34) is $w(\rho) = 1$. We conclude that

$$w(\rho, \theta, \varphi) = 1 + \frac{1}{84}\rho^2(\rho^2 - a^2)P_2^{(1)}(\cos\theta)\cos\varphi.$$

Remark 1. In the general case, when one solves the interior Dirichlet problem for the Laplace equation with the condition $u|_{\partial\Omega} = f(\theta, \varphi)$ (where Ω is the ball of radius a centered at the origin and $\partial\Omega$ is its boundary), one can write

$$f(\theta, \varphi) = \sum_{n=0}^{\infty}\sum_{m=0}^{n} a^n \left[A_{nm}\cos(m\varphi) + B_{nm}\sin(m\varphi)\right]P_n^{(m)}(\cos\theta),$$

where the coefficients A_{nm} and B_{nm} are given by the formulas

$$A_{nm} = \frac{\displaystyle\int_0^{2\pi}\int_0^{\pi} f(\theta, \varphi)P_n^{(m)}(\cos\theta)\cos(m\varphi)\sin\theta\,d\theta\,d\varphi}{\|Y_n^{(m)}\|^2 a^n}$$

and

$$B_{nm} = \frac{\displaystyle\int_0^{2\pi}\int_0^{\pi} f(\theta, \varphi)P_n^{(m)}(\cos\theta)\sin(m\varphi)\sin\theta\,d\theta\,d\varphi}{\|Y_n^{(m)}\|^2 a^n};$$

also,

$$\|Y_n^{(m)}\|^2 = \frac{2\pi\varepsilon_m}{2n+1}\frac{(n+m)!}{(n-m)!}, \quad \text{where} \quad \varepsilon_m = \begin{cases} 2, & \text{if } m = 0, \\ 1, & \text{if } m > 0. \end{cases}$$

Remark 2. The solution of the aforementioned interior Dirichlet problem for the Laplace equation at a point $(\rho_0, \theta_0, \varphi_0)$ admits the integral representation (Poisson integral)

$$u(\rho_0, \theta_0, \varphi_0) = \frac{a}{4\pi}\int_0^{2\pi}\int_0^{\pi} f(\theta, \varphi)\frac{a^2 - \rho^2}{(a^2 - 2a\rho_0\cos\gamma + \rho_0^2)^{3/2}}\sin\theta\,d\theta\,d\varphi,$$

where $\cos\gamma = \cos\theta\cos\theta_0 + \sin\theta\sin\theta_0\cos(\varphi - \varphi_0)$.

1.10. Boundary value problems for the Helmholtz equations

The Helmholtz equations $\Delta u + k^2 u = f$ and $\Delta u - k^2 u = f$, alongside with the Laplace and Poisson equations, are an important form of second-order elliptic equation. The homogeneous equation ($f = 0$), for example, arises naturally (in the multi-dimensional case) when the method of separation of variables is applied to hyperbolic and parabolic problems. Finding eigenvalues and eigenfunctions reduced to the solvability of the corresponding boundary value problem for a Helmholtz equation with $f \equiv 0$.

Example 1 [4, Ch. VII, no. 29(a)]. Find the natural oscillations of a membrane that has the shape of a annular sector ($a \le \rho \le b$, $0 \le \varphi \le \varphi_0$), with free boundary.

Solution. The problem is formulated mathematically as follows:

$$\begin{cases} \dfrac{1}{\rho}\dfrac{\partial}{\partial \rho}\left(\dfrac{\partial u}{\partial \rho}\right) + \dfrac{1}{\rho^2}\dfrac{\partial^2 u}{\partial \varphi^2} + \lambda u = 0, & a < \rho < b, \quad 0 < \varphi < \varphi_0, \\[2mm] \dfrac{\partial u}{\partial \rho}(a, \varphi) = \dfrac{\partial u}{\partial \rho}(b, \varphi) = 0, & 0 \le \varphi \le \varphi_0, \\[2mm] \dfrac{\partial u}{\partial \varphi}(\rho, 0) = \dfrac{\partial u}{\partial \varphi}(\rho, \varphi_0) = 0, & a \le \rho \le b. \end{cases} \tag{1.35}$$

We will seek the solution of this problem in the form

$$u(\rho, \varphi) = R(\rho)\Phi(\varphi).$$

Inserting this expression in equation (1.35) and separating the variables we obtain two ordinary differential equations:

$$(1) \quad \Phi'' + \nu \Phi = 0,$$

and

$$(2) \quad \rho\frac{d}{d\rho}(\rho R') + (\lambda \rho^2 - \nu)R = 0.$$

To determine ν we have the Sturm-Liouville problem

$$\begin{cases} \Phi'' + \nu \Phi = 0, & 0 < \varphi < \varphi_0, \\ \Phi'(0) = 0, \quad \Phi'(\varphi_0) = 0. \end{cases}$$

This yields $\nu_n = \left(\frac{\pi n}{\varphi_0}\right)^2$, $n = 0, 1, \ldots$, and $\Phi_n(\varphi) = \cos\left(\frac{\pi n}{\varphi_0}\varphi\right)$. The function $R(\rho)$ is obtained from the following boundary value problem for the Bessel equation

$$\begin{cases} \rho\dfrac{d}{d\rho}(\rho R') + (\lambda \rho^2 - \nu R) = 0, & a < \rho < b, \\ R'(a) = R'(b) = 0. \end{cases} \tag{1.36}$$

The general solution of equation (1.36) has the form

$$R(\rho) = C_1 J_{\frac{\pi n}{\varphi_0}}(\sqrt{\lambda}\rho) + C_2 N_{\frac{\pi n}{\varphi_0}}(\sqrt{\lambda}\rho),$$

where C_1 and C_2 are arbitrary constants and $N_{\frac{\pi n}{\varphi_0}}(\sqrt{\lambda}\rho)$ is the Bessel function of second kind. The values of λ are determined by means of the boundary conditions in (1.36); namely, they provide the system of equations

$$\begin{cases} C_1 J'_{\frac{\pi n}{\varphi_0}}(\sqrt{\lambda}a) + C_2 N'_{\frac{\pi n}{\varphi_0}}(\sqrt{\lambda}a) = 0, \\ C_1 J'_{\frac{\pi n}{\varphi_0}}(\sqrt{\lambda}b) + C_2 N'_{\frac{\pi n}{\varphi_0}}(\sqrt{\lambda}b) = 0. \end{cases}$$

This system has a nontrivial solution if and only if its determinant

$$\begin{vmatrix} J'_{\frac{\pi n}{\varphi_0}}(\sqrt{\lambda}a) & N'_{\frac{\pi n}{\varphi_0}}(\sqrt{\lambda}a) \\ J'_{\frac{\pi n}{\varphi_0}}(\sqrt{\lambda}b) & N'_{\frac{\pi n}{\varphi_0}}(\sqrt{\lambda}b) \end{vmatrix}$$

is equal to zero. In other words, $\lambda_{m,n} = \left[\mu_m^{(n)}\right]^2$, where $\mu_m^{(n)}$ are the roots of the equation

$$\frac{J'_{\frac{\pi n}{\varphi_0}}(\sqrt{\lambda}a)}{J'_{\frac{\pi n}{\varphi_0}}(\sqrt{\lambda}b)} = \frac{N'_{\frac{\pi n}{\varphi_0}}(\sqrt{\lambda}a)}{N'_{\frac{\pi n}{\varphi_0}}(\sqrt{\lambda}b)}.$$

We see that the radial function has the form

$$R_{m,n}(\rho) = J_{\frac{\pi n}{\varphi_0}}(\mu_m^{(n)}\rho) N'_{\frac{\pi n}{\varphi_0}}(\mu_m^{(n)}a) - J'_{\frac{\pi n}{\varphi_0}}(\mu_m^{(n)}a) N_{\frac{\pi n}{\varphi_0}}(\mu_m^{(n)}\rho).$$

Thus, the natural oscillations of our plate are described by the functions

$$u_{m,n}(\rho, \varphi) = R_{m,n}(\rho)\Phi_n(\varphi) =$$

$$= \left[J_{\frac{\pi n}{\varphi_0}}(\mu_m^{(n)}\rho) N'_{\frac{\pi n}{\varphi_0}}(\mu_m^{(n)}a) - J'_{\frac{\pi n}{\varphi_0}}(\mu_m^{(n)}a) N_{\frac{\pi n}{\varphi_0}}(\mu_m^{(n)}\rho) \right] \cos\left(\frac{\pi n}{\varphi_0}\varphi\right).$$

1.11. Boundary value problem for the Helmoltz equation in a cylinder

Example [4, Ch. VII, no. 10]. Find the steady distribution of the concentration of an unstable gas inside an infinite cylinder of circular section assuming that a constant concentration u_0 is maintained on the surface of the cylinder.

Solution. It is know that the problem of diffusion of an unstable gas that decomposes during the diffusion proces is described by the equation

$$\Delta u - \varkappa^2 u = 0 \qquad (\varkappa > 0).$$

Hence, in polar coordinates the problem is formulated as

$$
\begin{cases}
\dfrac{1}{\rho}\dfrac{\partial}{\partial\rho}\left(\dfrac{\partial u}{\partial\rho}\right) + \dfrac{1}{\rho^2}\dfrac{\partial^2 u}{\partial\varphi^2} - \varkappa^2 u = 0, & 0 < \rho < a, \quad 0 < \varphi < 2\pi, \\
u(a,\varphi) = u_0, & 0 \le \varphi \le 2\pi,
\end{cases}
\tag{1.37}
$$

where a denotes the radius of the cylinder.

Let us seek the solution in the form $u(\rho,\varphi) = R(\rho)\Phi(\varphi)$. Substituting this expression in equation (1.37) we obtain

$$
\frac{1}{\rho}\frac{d}{d\rho}(\rho R')\Phi + \frac{R}{\rho^2}\Phi'' - \varkappa^2 R\Phi = 0,
$$

or

$$
\frac{\rho\dfrac{1}{\rho}\dfrac{d}{d\rho}(\rho R')}{R} - \varkappa^2\rho^2 = -\frac{\Phi''}{\Phi} = \lambda.
$$

This yields two ordinary differrential equations:

$$
(1) \quad \Phi'' + \lambda\Phi = 0,
$$

and

$$
(2) \quad \rho\frac{d}{d\rho}(\rho R') - (\varkappa^2\rho^2 + \lambda)R = 0.
$$

From equation (1), by using the fact that $\Phi(\varphi) = \Phi(\varphi+\pi)$, we obtain $\lambda = n^2$ ($n = 0,1,2,\dots$) and $\Phi_n(\varphi) = A_n\cos(n\varphi) + B_n\sin(n\varphi)$, where A_n and B_n are arbitrary constants.

Further, equation (2) yields

$$
\rho^2 R'' + \rho R' - (\varkappa^2\rho^2 + n^2)R = 0.
$$

After the change of variables $x = \varkappa\rho$ we obtain the equation

$$
x^2\frac{d^2 R}{dx^2} + x\frac{dR}{dx} - (x^2 + n^2)R = 0.
$$

This is recognized as being the Bessel equation of imaginary argument of order n. Its general solution has the form

$$
R(x) = C_1 I_n(x) + C_2 K_n(x),
$$

where $I_n(x)$ and $K_n(x)$ are the cylindrical functions of imaginary argument of first and second kind, respectively. Clearly, we must put $C_2 = 0$ because the solution is required to be bounded on the axis of the cylinder ($K_n(x)$ has a logarithmic singularity as $x \to 0$). Returning to the original variable we write

$$R(\rho) = C I_n(\varkappa \rho),$$

where C is an arbitrary constant.

Thus,

$$u(\rho, \varphi) = \sum_{n=0}^{\infty} [A_n \cos(n\varphi) + B_n \sin(n\varphi)] I_n(\varkappa \rho),$$

where the constants A_n are determined from the boundary condition. Namely, we have

$$u(a, \varphi) = \sum_{n=0}^{\infty} [A_n \cos(n\varphi) + B_n \sin(n\varphi)] I_n(\varkappa \rho),$$

and since $u(a, \varphi) = u_0$, we see that $A_0 = u_0 / I_0(\varkappa a)$, while all the remaining terms of the series are equal to zero. Hence, the solution is

$$u(\rho, \varphi) = u_0 \frac{I_0(\varkappa \varphi)}{I_0(\varkappa a)}.$$

1.12. Boundary value problems for the Helmoltz equation in a disc

Example 1. Solve the following boundary value problem for the Helmholtz equation in a disc:

$$\begin{cases} \Delta u + k^2 u = 0, & 0 \le \varphi < 2\pi, \quad 0 < \rho < a, \\ u(a, \varphi) = f(\varphi), & 0 \le \varphi \le 2\pi; \end{cases}$$

here one assumes that k^2 is not equal to any of the eigenvalues λ of the homogeneous Dirichlet problem for the equation $\Delta u + \lambda u = 0$.

Solution. Using again separation of variables, we write $u(\rho, \varphi) = R(\rho)\Phi(\varphi)$, which upon substitution in the Helmholtz equation yields

$$\frac{1}{\rho}\frac{d}{d\rho}(\rho R') \cdot \Phi + R \frac{1}{\rho^2}\Phi'' + k^2 R\Phi = 0.$$

Hence,

$$\frac{\rho \dfrac{d}{d\rho}(\rho R')}{R} + k^2 \rho^2 = -\frac{\Phi''}{\Phi} = \lambda,$$

where λ is the separation constant.

The eigenvalues and corresponding eigenfunctions are obtained as the solutions of the already familiar problem

$$\begin{cases} \Phi'' + \lambda \Phi = 0, & -\infty < \varphi < \infty, \\ \Phi(\varphi) = \Phi(\varphi + 2\pi). \end{cases}$$

Hence, $\lambda = n^2$ and $\Phi_n(\varphi) = C_1 \cos(n\varphi) + C_2 \sin(n\varphi)$, $n = 0, 1, 2, \ldots$. Since

$$\frac{\rho \dfrac{d}{d\rho}(\rho R')}{R} + k^2 \rho^2 = \lambda,$$

we obtain the following equation for the determination of $R(\rho)$:

$$\rho \frac{d}{d\rho}(\rho R') + (k^2 \rho^2 - n^2)R = 0. \tag{1.38}$$

Denoting $x = k\rho$, we rewrite (1.38) in the form

$$x^2 \frac{d^2 R}{dx^2} + x \frac{dR}{dx} + (x^2 - n^2)R = 0.$$

This is the Bessel equation of order n and has the general solution

$$R(x) = C_1 J_n(x) + C_2 Y_n(x),$$

where $J_n(x)$ and $Y_n(x)$ are the nth order Bessel functions of the first and second kind, respectively, and C_1, C_2 are arbitrary constants.

Therefore, the solution of equation (1.38) has the form

$$R(\rho) = C_1 J_n(k\rho) + C_2 Y_n(k\rho).$$

Since $Y_n(k\rho) \to \infty$ as $\rho \to 0$ and we are interested in bounded solutions, we mus take $C_2 = 0$. Thus, $R_n(\rho) = J_n(k\rho)$ and the solution of our problem is represented as a series

$$u(\rho, \varphi) = \sum_{n=0}^{\infty} [A_n \cos(n\varphi) + B_n \sin(n\varphi)] J_n(k\rho). \tag{1.39}$$

The constants A_n and B_n are found from the boundary conditions. Setting $\rho = a$ in (1.39), we obtain

$$f(\varphi) = \sum_{n=0}^{\infty} [A_n \cos(n\varphi) + B_n \sin(n\varphi)] J_n(ka),$$

whence

$$A_n = \frac{1}{2\pi J_n(ka)} \int_0^{2\pi} f(\varphi) \cos(n\varphi) \, d\varphi, \qquad n = 0, 1, \ldots,$$

and

$$B_n = \frac{1}{2\pi J_n(ka)} \int_0^{2\pi} f(\varphi) \sin(n\varphi) \, d\varphi, \qquad n = 1, 2, \ldots.$$

In particular, if $f(\varphi) = A\sin(3\varphi)$, we have

$$B_3 = \frac{A}{J_3(ka)}, \quad A_n = 0, \quad n = 0, 1, \dots; \quad B_n = 0, \quad n \neq 3,$$

and the solution has the expression

$$u(\rho, \varphi) = \frac{A}{J_3(ka)}\, J_3(k\rho)\sin(3\varphi).$$

Problem 2. Solve the following Dirichlet problem for the Helmoltz equation:

$$\begin{cases} \Delta u + k^2 u = 0, & 0 \leq \varphi < 2\pi, \quad \rho > a, \\ u|_{\rho=a} = f(\varphi), & 0 \leq \varphi \leq 2\pi, \\ u_\rho + ik\rho = o(\rho^{-1/2}) & \text{as} \quad \rho \to \infty. \end{cases}$$

Solution. Here, as in the preceding example, we will use separation of variables to find the solution. The only difference is that in the present case, in order to make the solution unique, we must impose for $n = 2$ the radiation condition (Sommerfeld condition)

$$\frac{\partial u}{\partial \rho} + ik\rho = o(\rho^{-1/2}), \quad \rho \to \infty.$$

The solution of problem (1.38) takes now . the form

$$R(\rho) = C_1 H_n^{(1)}(k\rho) + C_2 H_n^{(2)}(k\rho),$$

where $H_n^{(1)}(x)$ and $H_n^{(2)}(x)$ are the Hankel functions of index n of the first and second kind, respectively. The behavior of the Hankel functions at infinity $\rho \to \infty$) is given by the asymptotic formulas

$$H_n^{(1)}(x) \sim \sqrt{\frac{2}{\pi x}}\, e^{i\left(x - \frac{\pi n}{2} - \frac{\pi}{4}\right)} \left[1 + O\left(\frac{1}{x}\right)\right],$$

and

$$H_n^{(2)}(x) \sim \sqrt{\frac{2}{\pi x}}\, e^{-i\left(x - \frac{\pi n}{2} - \frac{\pi}{4}\right)} \left[1 + O\left(\frac{1}{x}\right)\right].$$

It readily checked directly that the radiation condition is satsified by the function $H_n^{(2)}(k\rho)$.

We see that the solution of the above exterior Dirichlet problem for the the Helmholtz equation is given by the series

$$u(\rho, \varphi) = \sum_{n=0}^{\infty} [A_n \cos(n\varphi) + B_n \sin(\varphi)]\, H_n^{(2)}(k\rho),$$

where the coefficients A_n and B_n are given by the formulas

$$A_n = \frac{1}{2\pi H_n^{(2)}(ka)} \int_0^{2\pi} f(\varphi)\cos(n\varphi)\, d\varphi, \quad n = 0, 1, \dots$$

and

$$A_n = \frac{1}{2\pi H_n^{(2)}(ka)} \int_0^{2\pi} f(\varphi)\sin(n\varphi)\, d\varphi, \quad n = 1, 2, \dots$$

1.13. Boundary value problems for the Helmoltz equation in a ball

Let us consider several examples of solutions for the interior and exterior Dirichlet and Neumann boundary value problems in a ball.

Example 1 [4, Ch. VII, no. 12]. Find the steady distribution of the concentration of an unstable gas inside a sphere of radius a if on the surface of the sphere one maintains the concentration $u|_{\partial\Omega} = u_0 \cos\theta$ ($u_0 = \text{const}$).

Solution. The problem is formulated mathematically as follows:

$$\begin{cases} \dfrac{1}{\rho^2}\dfrac{\partial}{\partial\rho}\left(\rho^2\dfrac{\partial u}{\partial\rho}\right) + \dfrac{1}{\rho^2\sin\theta}\dfrac{\partial}{\partial\theta}\left(\sin\theta\dfrac{\partial u}{\partial\theta}\right) - \varkappa^2 u = 0, \\ 0 < \rho < a, \quad 0 < \theta < \pi, \\ u(a,\theta) = u_0\cos\theta, \quad 0 \le \theta \le \pi. \end{cases} \tag{1.40}$$

As before, let us seek the solution in the form

$$u(\rho,\theta) = R(\rho)T(\theta).$$

Substituting this expression in equation (1.40) we obtain

$$\frac{1}{\rho^2}\frac{\partial}{\partial\rho}(\rho^2 R')\cdot T + \frac{1}{\rho^2}\frac{1}{\sin\theta}R\cdot\frac{\partial}{\partial\theta}(\sin\theta\cdot T') - \varkappa^2 RT = 0$$

whence, upon dividing both sides by RT,

$$\frac{\dfrac{\partial}{\partial\rho}(\rho^2 R')}{R} - \varkappa^2\rho^2 = -\frac{\dfrac{1}{\sin\theta}\dfrac{\partial}{\partial\theta}(\sin\theta\cdot T')}{T} = \lambda.$$

This yields two ordinary differential equations:

$$(1) \qquad \frac{d}{d\rho}(\rho^2 R') - (\varkappa^2\rho^2 + \lambda)R = 0,$$

and

$$(2) \qquad \frac{1}{\sin\theta}\frac{\partial}{\partial\theta}(\sin\theta\cdot T') + \lambda T = 0.$$

Performing the change of variables $x = \cos\theta$ in equation (2) (and using the conditions $|T(0)| < \infty$, $|T(\pi)| < \infty$), we find the eigenvalues and eigenfunctions

$$\lambda_n = n(n+1), \qquad T_n(\theta) = P_n(\cos\theta), \quad n = 0,1\dots,$$

where $P_n(x)$ are the Legendre polynomials.

Equation (1) is readily reduced, via the substitution $v(\rho) = \sqrt{\bar{\rho}}R(\rho)$, to the form (for each n)

$$\rho^2 v'' + \rho v' - \left[(\varkappa\rho)^2 + \left(n + \frac{1}{2}\right)^2\right]v = 0.$$

The corresponding bounded solutions of this equations are

$$v_n(\rho) = CI_{n+1/2}(\varkappa\rho),$$

where $I_{n+1/2}(x)$ are the Bessel functions of half-integer order and imaginary argument. Then

$$R_n(\rho) = \frac{I_{n+1/2}(\varkappa\rho)}{\sqrt{\rho}}.$$

Therefore, the solution of our problem is given by the series

$$u(\rho, \theta) = \sum_{n=0}^{\infty} C_n \frac{I_{n+1/2}(\varkappa\rho)}{\sqrt{\rho}} P_n(\cos\theta),$$

where the constants C_n are determined from the boundary conditions. Specifically,

$$u_0 \cos\theta = \sum_{n=0}^{\infty} C_n \frac{I_{n+1/2}(\varkappa a)}{\sqrt{a}} P_n(\cos\theta).$$

This yields $C_1 = u_0\sqrt{a}/I_{3/2}(\varkappa a)$ (the remaining coefficients are equal to zero). Finally,

$$u(\rho, \theta) = u_0 \frac{\sqrt{a}}{\sqrt{\rho}} \frac{I_{3/2}(\varkappa\rho)}{I_{3/2}(\varkappa a)} \cos\theta.$$

Example 2 [6, Ch. IV, 18.51]. Solve the Neumann problem for the equation $\Delta u + k^2 u = 0$ in the interior as well as in the exterior of the sphere $\rho = R$, under the condition $\partial u/\partial n|_{\rho=R} = A$, where A is a constant.

Solution. (a) The interior Neumann problem can be written as follows:

$$\frac{1}{\rho^2} \frac{\partial}{\partial\rho}\left(\rho^2 \frac{\partial u}{\partial\rho}\right) + k^2 u = 0, \quad 0 < \rho < R, \quad 0 < \theta < \pi, \quad 0 \le \varphi < 2\pi, \quad (1.41)$$

$$\left.\frac{\partial u}{\partial n}\right|_{\rho=R} = A, \quad 0 \le \theta \le \pi, \quad 0 \le \varphi < 2\pi. \quad (1.42)$$

Since $\frac{1}{\rho^2} \frac{\partial}{\partial\rho}\left(\rho^2 \frac{\partial u}{\partial\rho}\right) = (\rho u)''$, equation (1.41) can be recast as

$$v'' + k^2 v = 0, \quad v(\rho) = \rho u(\rho).$$

The general solution of this equation is

$$v(\rho) = C_1 \cos(k\rho) + C_2 \sin(k\rho),$$

and consequently

$$u(\rho) = C_1 \frac{\cos(k\rho)}{\rho} + C_2 \frac{\sin(k\rho)}{\rho}.$$

Since the solution must be bounded at the center of the ball, we must put $C_1 = 0$, and so

$$u(\rho) = C \frac{\sin(k\rho)}{\rho}.$$

Now let us calculate the normal derivative:

$$\frac{\partial u}{\partial n} = \frac{\partial u}{\partial \rho} = C \frac{k \cos(k\rho) \cdot \rho - \sin(k\rho)}{\rho^2}.$$

Further, using the boundary condition (1.42) we obtain

$$C \frac{Rk \cos(kR) - \sin(kR)}{R^2} = A,$$

whence

$$C = \frac{AR^2}{kR \cos(kR) - \sin(kR)}.$$

We conclude that the solution of the interior problem has the form

$$u(\rho) = \frac{AR^2}{kR \cos(kR) - \sin(kR)} \cdot \frac{\sin(k\rho)}{\rho}.$$

(b) The exterior Neumann problem reads:

$$\frac{1}{\rho^2} \frac{\partial}{\partial \rho} \left(\rho^2 \frac{\partial u}{\partial \rho} \right) + k^2 u = 0, \quad \rho > R, \quad 0 \le \theta \le \pi, \quad 0 \le \varphi < 2\pi, \tag{1.43}$$

$$\left. \frac{\partial u}{\partial n} \right|_{\rho=R} = A, \quad 0 \le \theta \le \pi, \quad 0 \le \varphi < 2\pi, \tag{1.44}$$

$$u_\rho - iku = o(\rho^{-1}) \quad \text{as } \rho \to \infty. \tag{1.45}$$

As in item (a), equation (1.43) can be recast as

$$v'' + k^2 v = 0, \quad v(\rho) = \rho u(\rho).$$

The general solution of this equation is

$$v(\rho) = C_1 e^{ik\rho} + C_2 e^{-ik\rho}, \quad \rho \to \infty.$$

Therefore,

$$u(\rho) = C_1 \frac{e^{ik\rho}}{\rho} + C_2 \frac{e^{-ik\rho}}{\rho}.$$

Let us verify that the function $u_1(\rho) = \frac{e^{ik\rho}}{\rho}$ satisfies the Sommerfeld condition

$$\frac{\partial u_1}{\partial \rho} - iku_1 = o(\rho^{-1}) \qquad (\rho \to \infty),$$

i.e., that

$$\lim_{\rho \to \infty} \left[\rho \left(\frac{\partial u}{\partial \rho} - iku_1 \right) \right] = 0.$$

Indeed,

$$\rho \left(\frac{\partial u}{\partial \rho} - iku_1 \right) = \rho \left(iku_1 - \frac{1}{\rho} u_1 - iku_1 \right) = -\frac{e^{ik\rho}}{\rho} \quad \text{and} \quad \left| \frac{e^{ik\rho}}{\rho} \right| \leq \frac{1}{\rho}.$$

It follows that to pick a unique solution we must set $C_2 = 0$, and then $u(\rho) = C\frac{e^{ik\rho}}{\rho}$.
Now let us calculate the normal derivative:

$$\frac{\partial u}{\partial n} = -\frac{\partial u}{\partial \rho} = -\frac{\partial}{\partial \rho} \left(C \frac{e^{ik\rho}}{\rho} \right).$$

We have

$$\frac{\partial u}{\partial n} = \frac{Ce^{ik\rho}}{\rho^2} (1 - ik\rho),$$

and the boundary condition (1.44) yields

$$C = \frac{AR^2}{e^{ikR}(1 - ikR)}.$$

Thus, the solution of the exterior Neumann problem is given by the formula

$$u(\rho) = \frac{AR^2}{e^{ikR}(1 - ikR)} \frac{e^{ik\rho}}{\rho}.$$

Example 3 [6, Ch. V, 18.53]. Solve the Dirichlet problem for the equation $\Delta u - k^2 u = 0$ in the interior and in the exterior of the sphere of radius $\rho = R$ with the condition $u|_{\rho=R} = A$, where A is a constant.

Solution. (a) First let us solve the interior Dirichlet problem

$$\begin{cases} \Delta u - k^2 u = 0, & 0 < \rho < R, \quad 0 < \theta < \pi, \quad 0 \leq \varphi < 2\pi, \\ u|_{\rho=R} = A, & 0 \leq \theta \leq \pi, \quad 0 \leq \varphi < 2\pi. \end{cases}$$

By analogy with Example 2, we must solve the equation $v'' - k^2 v = 0$, where $v(\rho) = \rho u(\rho)$, which has the general solution

$$v(\rho) = C_1 \sinh(k\rho) + C_2 \cosh(k\rho).$$

Therefore,

$$u(\rho) = C_1 \frac{\sinh(k\rho)}{\rho} + C_2 \frac{\cosh(k\rho)}{\rho}.$$

where C_1 and C_2 are arbitrary constants.

Note that $\cosh(k\rho)/\rho \to \infty$ as $\rho \to 0$. Hence, we must put $C_2 = 0$, and the solution has the expression

$$u(\rho) = C \frac{\sinh(k\rho)}{\rho}.$$

The constant C is determined from the boundary condition $u(R) = A$, i.e., $C \sinh(kR)/R = A$, which yields $C = AR/\sinh(kR)$. We conclude that the solution of our problem is

$$u(\rho) = A \frac{R}{\rho} \frac{\sinh(k\rho)}{\sinh(kR)}.$$

(b) Now let us solve the exterior Dirichlet problem

$$\begin{cases} \Delta u - k^2 u = 0, & \rho > R, \quad 0 < \theta < \pi, \quad 0 \le \varphi < 2\pi, \\ u|_{\rho=R} = A, & 0 \le \theta \le \pi, \quad 0 \le \varphi < 2\pi, \\ u(\rho) \to 0 & \text{as } \rho \to \infty. \end{cases}$$

In this case

$$u(\rho) = C_1 \frac{e^{k\rho}}{\rho} + C_2 \frac{e^{-k\rho}}{\rho}.$$

Since the solution of the exterior problem must satisfy $u(\rho) \to 0$ when $\rho \to \infty$, we must put $C_1 = 0$. Therefore,

$$u(\rho) = C \frac{e^{-k\rho}}{\rho}.$$

The boundary condition $u|_{\rho=R} = A$ yields $C = AR/e^{-kR}$. We conclude that the solution of our problem is

$$u(\rho) = A \frac{R}{\rho} \frac{e^{-k\rho}}{e^{-kR}}.$$

1.14. Guided electromagnetic waves

In this section we will conisder problem connected with steady processes of propagation of electromagnetic waves in systems that have the property of producing conditions under which waves propagate essentially in a given direction. Such waves are know as *guided waves*, and the systems that guide them are called *waveguides*.

The basic tool that we will use to simplify the analysis of such problems is the representation of elecctromagnetic waves as a superposition of waves of several types.

Let us assume that the x_3-axis coincides with the direction of wave propagation. The electromagnetic field of the wave is described by six components, E_1, E_2, E_3, H_1, H_2, H_3, of the electric and magnetic field vectors. Let us write them in matrix form:

$$A = \begin{bmatrix} 0 & E_2 & 0 \\ H_1 & 0 & H_3 \end{bmatrix}, \qquad B = \begin{bmatrix} E_1 & 0 & E_3 \\ 0 & H_2 & 0 \end{bmatrix}.$$

It is clear that the electric field vector $(0, E_2, 0)$ is orthogonal to the direction of propagation of the wave, whereas the magnetic field vector $(H_1, 0, H_2)$ has a nonzero component along that direction. In the matrix B the vector $(E_1, 0, E_3)$ has a nonzero component along the x_3-axis, whereas the vector $(0, H_2, 0)$ is orthogonal to the x_3-axis. In connection with this circumstance the waves characterized by the matrix A are referred to as *transversally-electric* (TE-waves), while those characterized by the matrix B are referred to as *transversally-magnetic* (TM-waves).

It is convenient to consider that an electromagnetic wave is a TE-wave [resp., TM-wave] if $E_3 = 0$ [resp., $H_3 = 0$].

There exists also waves of a third type, characterized by the matrix

$$C = \begin{bmatrix} E_1 & E_2 & 0 \\ H_1 & H_2 & 0 \end{bmatrix}.$$

These are called *transversally- electromagnetic* waves, or TEM-waves.

Example. TM-waves in a waveguide of circular cross section.

Let us consider the propagation of TM-waves in an infinitely long cylinder of radius a. It is known that this problem is connected with the solvability of the following Dirichlet problem for the Helmholtz equation:

$$\begin{cases} \dfrac{1}{\rho}\dfrac{\partial}{\partial \rho}\left(\rho \dfrac{\partial E_3}{\partial \rho}\right) + \dfrac{1}{\rho^2}\dfrac{\partial^2 E_3}{\partial \varphi^2} + \delta^2 E_3 = 0, & 0 < \rho < a, \quad 0 \le \varphi < 2\pi, \\ E_3(a, \varphi) = 0, & 0 \le \varphi \le 2\pi, \end{cases}$$

where δ^2 is a real constant.

Separating the variables by means of the substitution $E_3 = R(\rho)\Phi(\varphi)$, we arrive at the equations

$$\begin{cases} \Phi'' + \lambda\Phi = 0, \\ \rho^2 R'' + \rho R' + (\delta^2\rho^2 - \lambda)R = 0, \end{cases} \tag{1.46}$$

where λ is the separation constant. Since $\Phi(\varphi) = \Phi(\varphi + 2\pi)$, it follows that $\lambda = n^2$, $n = 0, 1, 2, \ldots$

The change of variables $x = \delta\rho$ takes the second equation in (1.46) into the Bessel equation of order n in the new variable x. Since $|R(0)| < \infty$ and $R(a) = 0$, we have

$$R(\rho) = J_n(\delta_{nm}\rho).$$

Here $\delta_{nm} = \mu_{nm}/a$, where μ_{nm} is the mth positive root of the nth order Bessel function $J_n(x)$.

Therefore, our problem admits the following particular solutions:

$$E_{3,nm} = J_n(\delta_{nm}\rho) \left[A_{nm} \cos(n\varphi) + B_{nm} \sin(n\varphi) \right],$$

where A_{nm} and B_{nm} are arbitrary constants. Each of these solutions corresponds to a certain TM-wave, which can propagate without damping in the given waveguide.

Remark 1. The propagation of a TE-wave in an infinitely long cylinder is associated with the solvability of the Neumann problem for the Helmoltz equation:

$$
\begin{cases}
\dfrac{1}{\rho} \dfrac{\partial}{\partial \rho} \left(\rho \dfrac{\partial H_3}{\partial \rho} \right) + \dfrac{1}{\rho^2} \dfrac{\partial^2 H_3}{\partial \varphi^2} + \delta^2 H_3 = 0, \qquad 0 < \rho < a, \quad 0 \le \varphi < 2\pi, \\
\dfrac{\partial H_3}{\partial \vec{n}}(a, \varphi) = 0, \qquad 0 \le \varphi \le 2\pi;
\end{cases}
$$

here $\vec{n}$ is the unit outer normal of the cylindrical waveguide.

By analogy with the preceding example, we obtain

$$H_{3,nm} = J_n(\delta_{nm}\rho) \left[A_{nm} \cos(n\varphi) + B_{nm} \sin(n\varphi) \right],$$

where now $\delta_{nm} = \lambda_{nm}/a$, with λ_{nm} being the mth positive root of the equation $dJ_n(x)/dx = 0$, $n = 0, 1, 2, \ldots$

Remark. 2 If the component E_3 (or H_3) is known, then the other components of the electric and magnetic field vectors can be found by only one differentiation (this follows from the Maxwell equations for the electromagnetic field).

1.15. The method of conformal mappings
(for the solution of boundary value problems in the plane)

Methods of the theory of functions of a complex variables found wide and effective application in solving a large number of mathematical problems that arise in various fields of science. In particular, in many cases the application of complex functions yields simple methods for solving boundary value problems for the Laplace equation. This is the result of the intimate connection between analytic functions of a complex variable and harmonic functions of two real variables, and also of the invariance of the Laplace equation under conformal mappings.

Suppose one wants to solve the Laplace equation $u_{xx} + u_{yy} = 0$ with some boundary condition in a domain of complicated shape in the plane of the variables x, y. This boundary value problem can be transformed into a new boundary value problem, in which one is required to solve the Laplace quation $\tilde{u}_{\xi\xi} + \tilde{u}_{\eta\eta} = 0$ in a simpler domain of the variables ξ, η, and such that the second domain is obtained from the first one via a comformal mapping $\zeta = f(z)$, where $z = x+iy$, $\zeta = \xi+i\eta$, and $\tilde{u}(\zeta) = u(z)$ for $\zeta = f(z)$ (Figure 1.6).

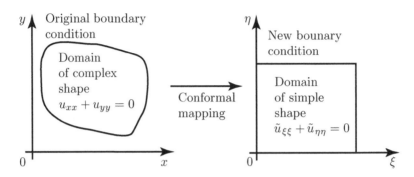

FIGURE 1.6.

Once the solution $\tilde{u}(\xi, \eta)$ of the Laplace equation in a simple domain (disc, half-space, rectangle) is found, it suffices to substitute in that solution the expressions $\xi = \xi(x, y)$, $\eta = \eta(x, y)$ in order to obtain the solution $u(x, y)$ of the original problem, expressed in the original variables.

Let us give several examples to show how to solve boundary value problems for the Laplace equation (in the plane) by means of conformal mappings.

Example 1 [6, Ch. V, 17.13(4)]. Find the solution of the equation $\Delta u = 0$ in the first quadrant $x > 0, y > 0$, with the boundary conditions $u|_{x=0} = 0$, $u|_{y=0} = \theta(x - 1)$, where $\theta(x) = 1$ if $x > 0$, $\theta(x) = 0$ if $x \leq 0$ is the Heaviside function.

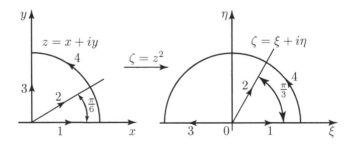

FIGURE 1.7.

Solution. Clearly, the function $\zeta = z^2$, defined in the first quadrant of the complex z-plane, maps this domain into the entire half-plane $\eta > 0$ of the complex ζ-plane (Figure 1.7), in such a manner that:

the positive x-semiaxis is mapped into the positive real ξ-semiaxis;
the positive y-semiaxis is mapped into the negative real ξ-semiaxis.

Thus, we arrive at the following conclusion:

Boundary value problem in the plane (x, y)

$$\begin{cases} \Delta u = 0, & x > 0, y > 0, \\ u|_{x=0} = 0, & y \geq 0, \\ u|_{y=0} = \theta(x - 1), & x \geq 0, \end{cases}$$

$\longrightarrow$

Boundary value problem in the plane (ξ, η)

$$\begin{cases} \Delta \tilde{u} = 0, & \eta > 0, \\ \tilde{u}|_{\eta=0} = \begin{cases} 1 & \text{if } \xi > 1, \\ 0 & \text{if } \xi < 1. \end{cases} \end{cases}$$

Notice also that from the equality $\zeta = z^2$, i.e., $\xi + i\eta = (x + iy)^2$, it follows that $\xi = x^2 - y^2$ and $\eta = 2xy$.

The solution of the Dirichlet problem in the (ξ, η)- plane is given by the Poisson integral

$$\tilde{u}(\xi, \eta) = \frac{\eta}{\pi} \int_{-\infty}^{\infty} \frac{\tilde{u}(t, 0)\, dt}{(t - \xi)^2 + \eta^2}.$$

Imposing the boundary condition on $\tilde{u}(\xi, 0)$, we obtain

$$\tilde{u}(\xi, \eta) = \frac{\eta}{\pi} \int_{-\infty}^{\infty} \frac{dt}{(t - \xi)^2 + \eta^2} = \frac{1}{\pi} \arctan \frac{t - \xi}{\eta} \Big|_1^{\infty} =$$

$$= \frac{1}{\pi} \left(\frac{\pi}{2} - \arctan \frac{1 - \xi}{\eta} \right) = \frac{1}{2} - \frac{1}{\pi} \arctan \frac{1 - \xi}{\eta}.$$

If we now write $x^2 - y^2$ instead of ξ and $2xy$ instead of η, we obtain the solution of the original problem in the form

$$u(x, y) = \frac{1}{2} - \frac{1}{\pi} \arctan \frac{y^2 - x^2 + 1}{2xy}.$$

Example 2 [6, Ch. V, 17.14(2)]. Find the solution of the Dirichlet problem for the equation $\Delta u = 0$ in the strip $0 < y < \pi$, with the boundary conditions $u|_{y=0} = \theta(x)$, $u|_{y=\pi} = 0$.

Solution. The complex function $\zeta = e^z$, defined in the strip $0 < y < \pi$, maps this strip into the entire half-plane $\eta > 0$ of the complex ζ-plane (Figure 1.8), in such a manner that:

the positive x-semiaxis is mapped into the positive ξ-semiaxis $[1, \infty)$;
the negative x-semiaxis is mapper into the interval $(0, 1)$;
the line $y = \pi$ is mapped into the negative ξ-semiaxis.

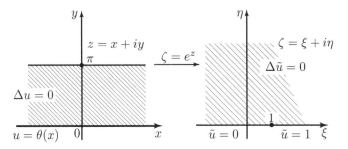

FIGURE 1.8.

Thus, we arrive at the following conclusion:

Boundary value problem in the plane (x, y)

$$\begin{cases} \Delta u = 0, & -\infty < x < \infty, \ 0 < y < \pi, \\ u|_{y=0} = \theta(x), & -\infty < x < \infty, \\ u|_{y=\pi} = 0, & -\infty < x < \infty, \end{cases}$$

$\longrightarrow$

Boundary value problem in the plane (ξ, η)

$$\begin{cases} \Delta \tilde{u} = 0, & \eta > 0, \\ \tilde{u}|_{\eta=0} = \begin{cases} 1 & \text{if } \xi > 1, \\ 0 & \text{if } \xi < 1. \end{cases} \end{cases}$$

Notice also that $\xi = e^x \cos y$ and $\eta = e^x \sin y$. As in Example 1, we have

$$\tilde{u}(\xi, \eta) = \frac{1}{2} - \frac{1}{\pi} \arctan \frac{1 - \xi}{\eta},$$

and so the solution has the form

$$u(x, y) = \frac{1}{2} - \frac{1}{\pi} \arctan \frac{e^{-x} - \cos y}{\sin y}.$$

Example 3 [6, Ch. V, 17.18]. Find the solution of the Dirichlet problem

$$\begin{cases} \Delta u = 0, & \operatorname{Re} z > 0, \quad |z - 5| > 3, \\ u|_{\operatorname{Re} z=0} = 0, & u|_{|z-5|=3} = 1. \end{cases}$$

Solution. First let us draw the domain D where we must solve the Dirichlet problem (Figure 1.9). It can be regarded as a eccentric annulus (indeed, a line is a circle of infinite radius). Let us find a conformal mapping of D onto a concentric annulus. To this purpose let us find two points that are simultaneously symmetric with respect to the line $\operatorname{Re} z = 0$ and with respect to the circle $|z - 5| = 3$. Clearly, such points must lie on the common perpendicular to the line and the circle, i.e., on the real axis. From the symmetry with respect to the line $\operatorname{Re} z = 0$ it follows that these are precisely the points $x_1 = a$ and $x_2 = -a$ with $a > 0$. The symmetry with

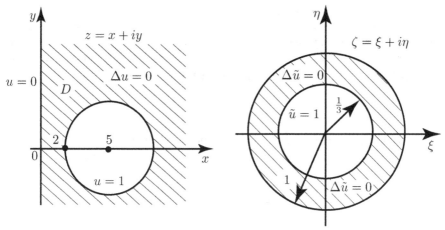

FIGURE 1.9.   FIGURE 1.10

respect to the circle $|z - 5| = 3$ translates into the equation $(5 + a)(5 - a) = 9$, which yields $a = 4$.

Let us show that the conformal transformation we are seeking is given by the linear-fractional function

$$\zeta = \frac{z - 4}{z + 4}. \tag{1.47}$$

Indeed, this mapping takes the line $\operatorname{Re} z = 0$ into a circle γ. Since symmetry must be preserved, the points $z_1 = 4$ and $z_2 = -4$ are taken into the points $\zeta = 0$ and $\zeta = \infty$, respectively, which are symmetric with respect to the circle γ. Hence, $\zeta = 0$ is the center of γ. Further, since the point $z = 0$ is taken into the point $\zeta = 1$, γ is the circle $|\zeta| = 1$ (Figure 1.10).

Now let us show than under the above mapping the circle $|z - 5| = 3$ goes into the circle $|\zeta| = 1/3$. Indeed, the linear-fractional-transformation (1.47) takes the circle $|z - 5| = 3$ into a circle, of radius $|\zeta| = |(2 - 4)/(2 + 4)| = 1/3$. We see that (1.47) maps the domain D conformally onto the concentric annulus $\frac{1}{3} < |\zeta| < 1$. We conclude that the given boundary value problem in the plane (x, y)

$$\begin{cases} \Delta u = 0, \quad \operatorname{Re} z > 0, \quad |z - 5| = 3, \\ u|_{\operatorname{Re} z = 0} = 0, \\ u|_{|z - 5| = 3} = 1, \end{cases}$$

is transformed into the following boundary value problem in the plane (ξ, η):

$$\Delta \tilde{u} = 0, \quad \frac{1}{3} < |\zeta| < 1, \tag{1.48}$$

$$\tilde{u}|_{|\zeta| = 1} = 0, \quad \tilde{u}|_{|\zeta| = \frac{1}{3}} = 1. \tag{1.49}$$

Let us solve the problem in the annulus $1/3 < |\zeta| < 1$ (in the plane (ξ, η)). Since the boundary conditions (1.49) do not depend on the polar angle φ, it is natural to assume that the solution $\widetilde{u}(\zeta)$ depends only on the variable ρ (here $\xi = \rho \cos \varphi$, $\eta = \rho \sin \varphi$). To find this solution, we rewrite the equation $\Delta \widetilde{u}$ in the form $\frac{\partial}{\partial \rho}\left(\rho \frac{\partial \widetilde{u}}{\partial \rho}\right) = 0$. The general solution of this equation is

$$\widetilde{u}(\zeta) = c_1 + c_2 \ln \rho,$$

where c_1 and c_2 are arbitrary constants. Imposing the conditions (1.49), we obtain $c_1 = 0$, $c_2 = -1/\ln 3$. Therefore,

$$\widetilde{u}(\zeta) = -\frac{1}{\ln 3} \ln |\zeta|, \quad \text{because } \rho = |\zeta|.$$

To find the solution of the original problem it suffices to return to the variable z, using (1.47), which finally yields

$$u(z) = \frac{1}{\ln 3} \ln \left|\frac{z+4}{z-4}\right|.$$

1.16. The Green function method

Definition of the Green function. Let us consider the boundary value problem

$$\begin{cases} \Delta u = f & \text{in the domain } \Omega, \\ \left(\alpha_1 u + \alpha_2 \dfrac{\partial u}{\partial n}\right) = g & \text{on the boundary } \partial\Omega. \end{cases} \tag{1.50}$$

We shall assume that the function $u(x)$ is continuous together with it first-order derivatives in the closed domain $\overline{\Omega} \subset \mathbf{R}^n$, bounded by a sufficiently smooth hypersurface $\partial\Omega$, and has second-order derivatives that are square integrable in Ω. Here $\vec{n}$ is the outward unit normal to $\partial\Omega$ and α_1, α_2 are given real numbers satisfying $\alpha_1^2 + \alpha_2^2 \neq 0$; $x = (x_1, \ldots, x_n)$.

The Green function method for solving such problems consists in the following. First we solve the auxiliary problem (see [1])

$$\begin{cases} \Delta G = -\delta(x, x_0), & x_0 \in \Omega, \\ \left(\alpha_1 G + \alpha_2 \dfrac{\partial G}{\partial n}\right)\bigg|_{\partial\Omega} = 0, \end{cases} \tag{1.51}$$

where $\delta = \delta(x, x_0)$ is the δ-function, which can formally be defined by the relations

$$\delta(x, x_0) = \begin{cases} 0, & \text{if } x \neq x_0, \\ \infty, & \text{if } x = x_0, \end{cases}, \qquad \int_\Omega \delta(x, x_0)dx = \begin{cases} 1, & \text{if } x_0 \in \Omega, \\ 0, & \text{if } x_0 \notin \Omega, \end{cases}$$

where $x_0 = (x_{01}, \ldots, x_{n0})$ (the notation dx is obvious). The main property of the δ-function is expressed by the equality

$$\int_\Omega \delta(x, x_0) f(x) dx = \begin{cases} f(x_0), & \text{if } x_0 \in \Omega, \\ 0, & \text{if } x_0 \notin \Omega, \end{cases}$$

where $f(x)$ is an arbitrary continuous function of the point x.

Definition. The solution of problem (1.51) is called the *Green function* of problem (1.50).

We will require that the Green function $G(x, x_0)$ be continuous (together with its first-order partial derivatives) everywhere in the closed domain $\overline{\Omega}$, except for the point x_0, at which $G(x, x_0)$ may have a singularity.

Once the function $G(x, x_0)$ is found, one can use it to easily find the solution of the original problem (1.50). To that end we will use the *second Green formula*

$$\int_\Omega (v\Delta u - u\Delta v) dx = \int_{\partial\Omega} \left(v\frac{\partial u}{\partial n} - u\frac{\partial v}{\partial n} \right) ds. \tag{1.52}$$

This formula is readily obtained from the Gauss-Ostrogradskiĭ formula

$$\int_{\partial\Omega} (\vec{a}, \vec{n}) ds = \int_\Omega \operatorname{div} \vec{a} \, dx$$

(where $\vec{a}$ is a vector field and $(\vec{a}, \vec{n})$ denotes the scalar product of the vectors $\vec{a}$ and $\vec{n}$) if one puts successively $\vec{a} = v\nabla u$ and $\vec{a} = u\nabla v$ and subtract the results from one another. Indeed, we have

$$\int_{\partial\Omega} v(\nabla u, \vec{n}) ds = \int_\Omega \operatorname{div}(v\nabla u) dx, \tag{1.53}$$

and

$$\int_{\partial\Omega} u(\nabla v, \vec{n}) ds = \int_\Omega \operatorname{div}(u\nabla v) dx. \tag{1.54}$$

Since $(\nabla u, \vec{n}) = \partial u/\partial n$, $(\nabla v, \vec{n}) = \partial v/\partial n$, $\operatorname{div}(v\nabla u) = (\nabla u, \nabla v) + v\Delta u$ and $\operatorname{div}(u\nabla u) = (\nabla u, \nabla v) + u\Delta v$, subtracting (1.54) from (1.53) we get the second Green formula.

Now let us put $v = G$ in (1.52). Then, since $\Delta u = f(x)$ and $\Delta G = -\delta(x, x_0)$, we obtain

$$\int_\Omega G(x, x_0) f(x) dx + \int_\Omega u(x)\delta(x, x_0) dx = \int_{\partial\Omega} \left(G\frac{\partial u}{\partial n} - u\frac{\partial G}{\partial n} \right) ds.$$

But, by the main property of the δ-function,

$$\int_\Omega u(x)\delta(x, x_0) dx = u(x_0),$$

and so the last equality yields

$$u(x_0) = \int_{\partial\Omega} \left(G\frac{\partial u}{\partial n} - u\frac{\partial G}{\partial n} \right) ds - \int_{\Omega} G(x, x_0)f(x)dx.$$

From this formula we obtain:

(a) the solution of the Dirichlet problem for

$$\alpha_1 = 1, \quad \alpha_2 = 0, \quad G|_{\partial\Omega} = 0, \quad u|_{\partial\Omega} = g$$

in the form

$$u(x_0) = -\int_{\partial\Omega} g\frac{\partial G}{\partial n}ds - \int_{\Omega} G(x, x_0)f(x)dx;$$

(b) the solution of the Neumann problem for

$$\alpha_1 = 0, \quad \alpha_2 = 1, \quad \frac{\partial G}{\partial n}\Big|_{\partial\Omega} = 0, \quad \frac{\partial u}{\partial n}\Big|_{\partial\Omega} = g$$

in the form

$$u(x_0) = \int_{\partial\Omega} Gg\,ds - \int_{\partial\Omega} G(x, x_0)f(x)dx.$$

Remark 1. The integral

$$\int_{\partial\Omega} G(x, x_0)f(x)dx$$

admits the following physical interpretation: the right-hand side of the equation is regarded as an external action on the system and is decomposed into a continual contribution of source distributed over the domain Ω. Then one finds the response of the system to each such source and one sums all these responses.

Construction of the Green function. One of the methods for constructing the Green function is the *reflection method*. For example, the Green function for the Poisson equation in the case of the half-space $(z > 0)$ has the form

$$G(M, M_0) = \frac{1}{4\pi R_{MM_0}} - \frac{1}{4\pi R_{MM_1}},$$

where R_{AB} denotes the distance between the points A and B, $M_0(x_0, y_0, z_0)$ is a point lying in the uper half-plane $z > 0$, $M_1(x_0, y_0, -z_0)$ is the point symmetric to $M_0(x_0, y_0, z_0)$ with respect to the plane $z = 0$, and $M(x, y, z)$ is an arbitrary point of the half-plane $z > 0$.

Physically the Green function can be interpreted as the potential of the field produced by point-like charges placed at the point M_0 (over the grounded plane $z = 0$) and the point M_1 (Figure 1.11).

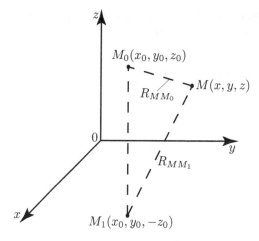

FIGURE 1.11. The potential at the point $M(x, y, z)$ equals
$$G(M, M_0) = \frac{1}{4\pi R_{MM_0}} - \frac{1}{4\pi R_{MM_1}}$$
($z = 0$ is a grounded conducting plane)

Notice that in the case of a half-plane ($y > 0$) the Green function has the form (Figure 1.12)

$$G(M, M_0) = \frac{1}{2\pi} \ln \frac{1}{R_{MM_0}} - \frac{1}{2\pi} \ln \frac{1}{R_{MM_1}}.$$

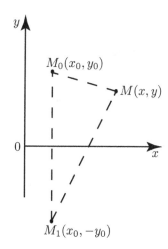

FIGURE 1.12. The potential at the point $M(x, y)$ equals
$$G(M, M_0) = \frac{1}{2\pi} \ln \frac{1}{R_{MM_0}} - \frac{1}{2\pi} \ln \frac{1}{R_{MM_1}}$$

Examples of problems solved by means of the Green function. Suppose we want to solve the Dirichlet problem for the Laplace equation in a half-plane

$$\begin{cases} \Delta u = 0, & -\infty < x < \infty, \quad y > 0, \\ u(x,0) = f(x), & -\infty < x < \infty. \end{cases}$$

The solution of this problem is

$$u(x,y) = -\frac{y}{\pi} \int_{-\infty}^{\infty} f(s) \left. \frac{\partial G}{\partial t} \right|_{t=0} ds$$

(we put $M_0 = M_0(x,y)$, $M = M(s,t)$), where

$$G(x,y;s,t) = \frac{1}{2\pi} \frac{1}{\sqrt{(x-s)^2 + (y-t)^2}} - \frac{1}{2\pi} \frac{1}{\sqrt{(x-s)^2 + (y+t)^2}}.$$

Calculating $\partial G/\partial t|_{t=0}$, we obtain

$$u(x,y) = \frac{y}{\pi} \int_{-\infty}^{\infty} \frac{f(s)}{(s-x)^2 + y^2} \, ds. \tag{1.55}$$

Example 1 [3, no. 244]. Find a function $u(x,y)$, harmonic in the half-plane $y > 0$, if it is known that

$$u(x,0) = \frac{x}{x^2 + 1}.$$

Solution. We must calculate the integral

$$u(x,y) = \frac{y}{\pi} \int_{-\infty}^{\infty} \frac{s}{(1+s^2)[(s-x)^2 + y^2]} \, ds.$$

Apparently, the easiest way to do this is to use the method of residues, namely, the following formula:

$$\int_{-\infty}^{\infty} \frac{s}{(1+s^2)[(s-x)^2 + y^2]} \, ds = 2\pi i [\operatorname{res}[f(z)]_{z=i} + \operatorname{res}[f(z)]_{z=x+iy},$$

where $f(z) = z/\left((1+z^2)[(z-x)^2 + y^2]\right)$.

Since

$$\operatorname{res}[f]_{z=i} = \frac{1}{2[(i-x)^2 + y^2]}, \qquad \operatorname{res}[f(z)]_{z=x+iy} = \frac{x+iy}{2iy[1 + (x+iy)^2]},$$

it follows that

$$\frac{y}{\pi} \int_{-\infty}^{\infty} \frac{s}{(1+s^2)[(s-x)^2 + y^2]} \, ds = \frac{iy}{[(i-x)^2 + y^2]} + \frac{x+iy}{[1 + (x+iy)^2]} =$$

$$= \frac{iy}{[(i-x)+iy][(i-x)-iy]} + \frac{x+iy}{(x+iy-1)(x+iy+1)} =$$

$$= \frac{1}{2}\left[\frac{1}{i-x-iy} - \frac{1}{i-x+iy}\right] + \frac{1}{2}\left[\frac{1}{x+iy-1} + \frac{1}{x+iy+1}\right] =$$

$$= \frac{1}{2}\left[\frac{1}{i(1-y)-x} - \frac{1}{i(1+y)-x} + \frac{1}{i(y-1)+x} + \frac{1}{i(1+y)+x}\right] =$$

$$= \frac{1}{2}\left[\frac{1}{i(1+y)+x} - \frac{1}{i(1+y)-x}\right] =$$

$$= \frac{1}{2}\left[\frac{x-i(1+y)}{x^2+(1+y)^2} + \frac{x+i(1+y)}{x^2+(1+y)^2}\right] = \frac{x}{x^2+(1+y)^2}.$$

Therefore, the solution of the problem is given by

$$u(x,y) = \frac{x}{x^2+(1+y)^2}.$$

Remark 2. The solution of the problem considered above,

$$\begin{cases} \Delta u = 0, & -\infty < x < \infty, \quad y > 0, \\ u|_{y=0} = \dfrac{x}{x^2+1}, \end{cases}$$

can also be obtained without resorting to the Green function.

Indeed, one can use the fact that the function $u = \frac{1}{2\pi} \ln \frac{1}{\sqrt{x^2+(y+a)^2}}$, where $a \geq 0$, is a solution of the Laplace equation in the upper half-plane $y > 0$, i.e.,

$$\Delta \ln \frac{1}{\sqrt{x^2+(y+a)^2}} = 0.$$

Differentiating this equality with respect to x, we obtain

$$\frac{\partial}{\partial x}\Delta \ln \frac{1}{\sqrt{x^2+(y+a)^2}} = 0, \quad \text{or} \quad \Delta \frac{\partial}{\partial x} \ln \frac{1}{\sqrt{x^2+(y+a)^2}} = 0,$$

i.e., $\Delta(x/r^2) = 0$, where $r = \sqrt{x^2+y^2}$.

Thus, the function $u = x/[x^2+(y+a)^2]$ is harmonic in the upper half-plane. Imposing the boundary condition, we conclude that the solution of our Dirichlet problem is the function

$$u(x,y) = \frac{x}{x^2+(y+1)^2}.$$

Example 2 [6, Ch. V, 17.4(2)]. Find the solution of the Dirichlet problem

$$\begin{cases} \Delta u = 0, & -\infty < x, y < \infty, \quad z > 0, \\ u|_{z=0} = \cos x \cos y, & -\infty < x, y < \infty. \end{cases}$$

Solution. It is known that the harmonic function we are asked to find is given by formula

$$u(x, y, z) = \frac{z}{2\pi} \int\!\!\!\int\limits_{-\infty}^{\infty} \frac{\cos\xi \cos\eta \, d\xi \, d\eta}{[(\xi - x)^2 + (\eta - y)^2 + z^2]^{3/2}} \, .$$

To calculate this integral we will make change the variables $\xi - x = u$, $\eta - y = v$, the Jacobian of which is 1. We obtain

$$u(x, y, z) = \frac{z}{2\pi} \int\!\!\!\int\limits_{-\infty}^{\infty} \frac{\cos(u + x) \cos(v + y) \, du \, dv}{(u^2 + v^2 + z^2)^{3/2}} =$$

$$= \frac{z}{2\pi} \int\!\!\!\int\limits_{-\infty}^{\infty} \frac{(\cos u \cos x - \sin u \sin x)(\cos v \cos y - \sin v \sin y) \, du \, dv}{(u^2 + v^2 + z^2)^{3/2}} =$$

$$= \frac{z}{2\pi} \cos x \cos y \int\!\!\!\int\limits_{-\infty}^{\infty} \frac{\cos u \cos v \, du \, dv}{(u^2 + v^2 + z^2)^{3/2}}$$

because the other three integrals vanish thanks to the fact that their integrands are odd functions.

Now let us calculate the integral

$$J = \int\!\!\!\int\limits_{-\infty}^{\infty} \frac{\cos u \cos v \, du \, dv}{(u^2 + v^2 + z^2)^{3/2}} =$$

$$= \int\!\!\!\int\limits_{-\infty}^{\infty} \frac{[\cos(u + v) + \sin u \sin v] du \, dv}{(u^2 + v^2 + z^2)^{3/2}} = \int\!\!\!\int\limits_{-\infty}^{\infty} \frac{\cos(u + v) \, du \, dv}{(u^2 + v^2 + z^2)^{3/2}},$$

because the other integral is equal to zero.

Let us make the change of variables

$$p = \frac{1}{\sqrt{2}}(u + v), \quad q = \frac{1}{\sqrt{2}}(u - v),$$

which correspond to a counter-clockwise rotation of the plane by $45°$. Then we have

$$\int\!\!\!\int\limits_{-\infty}^{\infty} \frac{\cos(\sqrt{2}p) \, dp \, dq}{(p^2 + q^2 + z^2)^{3/2}} = \int_{-\infty}^{\infty} \cos(\sqrt{2}p) \, dp \int_{-\infty}^{\infty} \frac{dq}{(p^2 + q^2 + z^2)^{3/2}} \, .$$

But the substitution $q = \sqrt{p^2 + z^2}\tan t$ transforms the integral

$$J_1 = \int_{-\infty}^{\infty} \frac{dq}{(p^2 + q^2 + z^2)^{3/2}}$$

into

$$J_1 = \int_{-\pi/2}^{\pi/2} \frac{|\cos t|}{p^2 + z^2}\,dt = \frac{2}{p^2 + z^2}\,.$$

Finally, the resulting integral

$$J = 2\int_{-\infty}^{\infty} \frac{\cos(\sqrt{2}p)\,dp}{p^2 + z^2}$$

is calculated using the Cauchy residue theorem as follows:

$$J = 2\operatorname{Re}\int_{-\infty}^{\infty} \frac{e^{i\sqrt{2}p}dp}{p^2 + z^2} = 4\pi i\operatorname{res}\left[\frac{e^{i\sqrt{2}p}}{p^2 + z^2}\right]_{p=zi} =$$

$$= 4\pi i\,\frac{e^{-\sqrt{2}z}}{2zi} = \frac{2\pi}{z}\,e^{-\sqrt{2}z}.$$

Therefore the solution is

$$u(x, y, z) = e^{-\sqrt{2}z}\cos x\cos y.$$

Remark 3. Since $y/[(t - x)^2 + y^2] = \operatorname{Re}[1/(i(t - z))]$, where $z = x + iy$, the Poisson formula (1.55) can be recast as

$$u(z) = \operatorname{Re}\frac{1}{\pi i}\int_{-\infty}^{\infty} \frac{u(t)\,dt}{t - z}\,. \qquad (1.56)$$

Now let us consider the Dirichlet problem for the Laplace equation in the half-plane $\operatorname{Im} z > 0$ (i.e., for $y > 0$):

$$\begin{cases} \Delta u = 0, & -\infty < x < \infty, \quad y > 0, \\ u|_{y=0} = R(x), & -\infty < x < \infty, \end{cases}$$

where the rational function $R(z)$ is real, has no poles on the real axis, and $R(z) \to 0$ when $z \to \infty$. By (1.56), the solution of this problem is the function

$$u(z) = \operatorname{Re}\frac{1}{\pi i}\int_{-\infty}^{\infty} \frac{R(t)\,dt}{t - z}\,.$$

This integral can be calculated by using Cauchy's residue theorem:

$$u(z) = -2\operatorname{Re} \sum_{\operatorname{Im} \zeta_k < 0} \operatorname{res} \left[\frac{R(\zeta)}{\zeta - z} \right]_{\zeta = \zeta_k}, \tag{1.57}$$

where the residues are taken for all poles of the function $R(z)$ in the lower half-plane $\operatorname{Im} z < 0$.

Example 3. Solve the Dirichlet problem

$$\begin{cases} \Delta u = 0, & -\infty < x < \infty, \quad y > 0, \\ u|_{y=0} = \dfrac{k}{1 + x^2}, & k = \text{const}, \quad -\infty < x < \infty. \end{cases}$$

Solution. Using formula (1.57), we have

$$u(z) = -2\operatorname{Re} \operatorname{res} \left[\frac{k}{(1 + \zeta^2)(\zeta - z)} \right]_{\zeta = -i} = -2\operatorname{Re} \frac{k}{2i(z + i)} = \frac{k(y + 1)}{x^2 + (y + 1)^2}.$$

1.17. Other methods

In this section we will consider methods for solving boundary value problems for the biharmonic equation and the equations $\Delta^2 u = f$, as well as boundary value problemd for the Laplace and Poisson equations (without employing the Green function).

Biharmonic equation.

Example 1. Solve the following boundary value problem in the disc $\{(\rho, \varphi) : 0 \leq \rho \leq a, 0 \leq \varphi < 2\pi\}$:

$$\begin{cases} \Delta^2 u = 0 & \text{in the disc,} \\ u|_{\rho=a} = 0, \quad \dfrac{\partial u}{\partial n} \bigg|_{\rho=a} = A \cos \varphi & \text{on the boundary of the disc.} \end{cases}$$

Here $\vec{n}$ is the unit outward normal to the boundary of the disc.

Solution. What we have is the Dirichlet problem for the biharmonic equation. It is known that it has a unique solution, given by the formula

$$u(\rho, \varphi) = \frac{1}{2\pi a} (\rho^2 - a^2)^2 \left[\frac{1}{2} \int_0^{2\pi} \frac{-g \, d\alpha}{\rho^2 + a^2 - 2a\rho \cos(\varphi - \alpha)} + \right.$$

$$\left. + \int_0^{2\pi} \frac{f[a - \rho \cos(\varphi - \alpha)] \, d\alpha}{[\rho^2 + a^2 - 2a\rho \cos(\varphi - \alpha)]^2} \right] \tag{1.58}$$

(here $f = u|_{\rho=a}$ and $g = \partial u/\partial n|_{\rho=a}$.)

In our case $f = 0$, $g = A \cos \varphi$, and so the solution is

$$u(\rho, \varphi) = \frac{1}{2\pi a} (\rho^2 - a^2)^2 \left(-\frac{1}{2} \right) \int_0^{2\pi} \frac{A \cos \alpha \, d\alpha}{\rho^2 + a^2 - 2a\rho \cos(\varphi - \alpha)} =$$

$$= -\frac{A(\rho^2 - a^2)^2}{4\pi a} \int_0^{2\pi} \frac{\cos \alpha \, d\alpha}{\rho^2 + a^2 - 2a\rho \cos(\varphi - \alpha)},$$

or

$$u(\rho, \varphi) = -\frac{A}{2a}(a^2 - \rho^2) \cdot \frac{1}{2\pi} \int_0^{2\pi} \frac{(a^2 - \rho^2) \cos \alpha \, d\alpha}{\rho^2 + a^2 - 2a\rho \cos(\varphi - \alpha)}.$$

To compute the last integral we remark that it yields a solution of the following Dirichlet problem for the Laplace equation in the disc:

$$\begin{cases} \Delta v = 0 & \text{in the disc,} \\ v|_{\rho=a} = \cos \varphi & \text{on the boundary of the disc.} \end{cases}$$

But the solution of this problem is clearly $v = \frac{\rho}{a} \cos \varphi$. Then, by the uniqueness of the solution of the Dirichlet problem for the Laplace equation, we have the identity

$$\frac{1}{2\pi} \int_0^{2\pi} \frac{(a^2 - \rho^2) \cos \alpha \, d\alpha}{\rho^2 + a^2 - 2a\rho \cos(\varphi - \alpha)} = \frac{\rho}{a} \cos \varphi.$$

Therefore, the solution of our problem is

$$u(\rho, \varphi) = \frac{A\rho(\rho^2 - a^2)}{2a^2} \cos \varphi.$$

Example 2. Solve the following boundary value problem in the disc $\{(\rho, \varphi) : 0 \le \rho \le a, 0 \le \varphi < 2\pi\}$:

$$\begin{cases} \Delta^2 u = 1 & \text{in the disc,} \\ u|_{\rho=a} = 0, \quad \dfrac{\partial u}{\partial n}\bigg|_{\rho=a} = 0 & \text{on the boundary of the disc.} \end{cases}$$

Solution. One can consider that the solution of the problem depends only on the variable ρ, i.e., $u = u(\rho)$. Next, let us remark that

$$\Delta^2 u = \left(\frac{\partial^4}{\partial \rho^4} + \frac{2}{\rho} \frac{\partial^3}{\partial \rho^3} - \frac{1}{\rho^2} \frac{\partial^2}{\partial \rho^2} \right) u + \frac{1}{\rho^3} \frac{\partial u}{\partial \rho},$$

and so we obtain a boundary value problem for an ordinary differential equation:

$$\begin{cases} \dfrac{\partial^4 u}{\partial \rho^4} + \dfrac{2}{\rho} \dfrac{\partial^3 u}{\partial \rho^3} - \dfrac{1}{\rho^2} \dfrac{\partial^2 u}{\partial \rho^2} + \dfrac{1}{\rho^3} \dfrac{\partial u}{\partial \rho} = 1, & 0 < \rho < a, \qquad (1.61) \\ u|_{\rho=a} = 0, \quad \dfrac{du}{d\rho}\bigg|_{\rho=a} = 0. & (1.62) \end{cases}$$

Equation (1.61) can be rewritten in the form

$$\rho^3 u'''' + 2\rho^2 u''' - \rho u'' + u' = \rho^3.$$

Let us denote $v = \frac{du}{d\rho}$. Then we obtain a third-order equation for the function $v = v(\rho)$:

$$\rho^3 v''' + 2\rho^2 v'' - \rho v' + v = \rho^3,$$

which is recongnized to be the well-known Euler equation. Its general solution is given by the function

$$v(\rho) = C_1 \rho^{-1} + C_2 \rho \ln\rho + A\rho + \frac{1}{16}\rho^3.$$

We must take $C_1 = 0$ and $C_2 = 0$, because otherwise the function $v'(\rho)$ would become infinite at the center of the disc (i.e., when $\rho \to 0$). Therefore, $u' = A\rho + \frac{1}{16}\rho^3$, and so

$$u(\rho) = \frac{A\rho^2}{2} + \frac{1}{64}\rho^4 + B.$$

The constants A and B are found from the boundary conditions (1.62). We conclude that the solution is

$$u(\rho) = \frac{1}{64}\left(a^2 - \rho^2\right)^2,$$

or

$$\dot u(\rho) = \frac{a^4}{64}\left[1 - \left(\frac{\rho}{a}\right)^2\right]^2.$$

Example 3. Solve the following boundary value problem in the half-plane $\{(x,y) : -\infty < x < \infty, y > 0\}$:

$$\Delta^2 u = e^{-2y}\sin x \qquad \text{in the half-plane,} \tag{1.63}$$

$$u|_{y=0} = 0, \quad \frac{\partial u}{\partial y}\bigg|_{y=0} = 0 \qquad \text{on the boundary of the half-plane.} \tag{1.64}$$

Let us rewrite the equation (1.63) in the form

$$\frac{\partial^4 u}{\partial x^4} + 2\frac{\partial^4 u}{\partial x^2 \partial y^2} + \frac{\partial^4 u}{\partial y^4} = e^{-2y}\sin x. \tag{1.65}$$

We shall seek for $u(x,y)$ in the form $u(x,y) = f(y)\sin x$, where the function $f(y)$ is subject to determination. Substituting this expression in equation (1.65) we obtain

$$f(y)\sin x + 2f''(y)(-\sin x) + f^{(iv)}(y)\sin x = e^{-2y}\sin x,$$

whence
$$f^{(iv)} - 2f'' + f = e^{-2y}. \tag{1.66}$$

The general solution of equation (1.66) has the form

$$f(y) = C_1 e^y + C_2 y e^y + C_3 e^{-y} + C_4 y e^{-y} + \frac{1}{9} e^{-2y}.$$

The constants C_1 and C_2 are equal to zero: otherwise, $f(y) \to \infty$ as $y \to \infty$. Hence,

$$f(y) = C_3 e^{-y} + C_4 y e^{-y} + \frac{1}{9} e^{-2y}.$$

The constants C_3 and C_4 are found from the boundary conditions (1.64), which translate into $f(0) = 0$ and $f'(0) = 0$. We have

$$f(y) = -\frac{1}{9} e^{-y} + \frac{1}{9} y e^{-y} + \frac{1}{9} e^{-2y}.$$

Thus, the solution of our problem is

$$u(x, y) = \frac{1}{9} (e^{-2y} - e^{-y} + y e^{-y}) \sin x.$$

The Laplace and Poisson equations.

Example 4. Solve the following boundary value problem in the half-space $\{(x, y, z) : -\infty < x, y < \infty, z > 0\}$:

$$\begin{cases} \Delta u = z e^{-z} \sin x \sin y & \text{in the half-space,} \\ u|_{z=0} = 0. \end{cases}$$

Solution. We will seek the function $u = u(x, y, z)$ in the form

$$u = f(z) \sin x \sin y,$$

where the function $f(z)$ needs to be determined. Then we get

$$\Delta u = \frac{\partial^2 u}{\partial x^2} + \frac{\partial^2 u}{\partial y^2} + \frac{\partial^2 u}{\partial y^2} = -f \sin x \sin y - f \sin x \sin y + f'' \sin x \sin y,$$

and so our equation becomes

$$-2f \sin x \sin y + f'' \sin x \sin y = z e^{-z} \sin x \sin y.$$

Hence, to find $f(z)$ we must solve the ordinary differential equation

$$f'' - 2f = z e^{-z}.$$

Its general solution is

$$f(z) = C_1 e^{\sqrt{2}z} + C_2 e^{-\sqrt{2}z} + e^{-z}(2 - z).$$

The constants C_1 and C_2 are found from the boundary conditions. First notice that $C_1 = 0$, because otherwise $f(z) \to \infty$ when $z \to \infty$. Therefore,

$$f(z) = C_2 e^{-\sqrt{2}z} + e^{-z}(2 - z).$$

Putting here $z = 0$ we find $f(0) = C_2 + 2$, and since $f(0) = 0$, it follows that $C_2 = -2$.

Therefore, the solution of our problem is

$$u(x, y, z) = [e^{-z}(2 - z) - 2e^{-\sqrt{2}z}]\sin x \sin y.$$

Example 3. Solve the boundary problem in the half-space $\{(x, y, z) : -\infty < x, y < \infty, z > 0\}$

$$\begin{cases} \Delta u = 0 & \text{in the half-space,} \\ u|_{z=0} = \dfrac{x^2 + y^2 - 2}{(1 + x^2 + y^2)^{5/2}}. \end{cases}$$

Solution. Notice that the function

$$u(x, y, z) = \frac{1}{\sqrt{x^2 + y^2 + (z + 1)^2}}$$

satisfies the Laplace equation in the whole half-space $z > 0$ (is a fundamental solution), i.e.,

$$\Delta \frac{1}{\sqrt{x^2 + y^2 + (z + 1)^2}} = 0.$$

Now let us differentiate both sides of this equality with respect to z. We get

$$\Delta \frac{z + 1}{[x^2 + y^2 + (z + 1)^2]^{3/2}} = 0.$$

Differentiating one more time with respect to z we have

$$\Delta \frac{x^2 + y^2 - 2(z + 1)^2}{[x^2 + y^2 + (z + 1)^2]^{5/2}} = 0.$$

This suggests to consider the function

$$u(x, y, z) = \frac{x^2 + y^2 - 2(z + 1)^2}{[x^2 + y^2 + (z + 1)^2]^{5/2}}.$$

This function is harmonic in the whole half-space $z > 0$ (since $\Delta u = 0$, as we just showed), and for $z = 0$ we have

$$u|_{z=0} = \frac{x^2 + y^2 - 2}{(1 + x^2 + y^2)^{5/2}},$$

which proves that $u(x, y, z)$ is the sought solution.

1.18. Problems for independent study

1. Find the distribution of the potential of an infinitely long $(-\infty < z < \infty)$ long cylindrical capacitor if its interior plate $\rho = a$ [resp., exterior plate $\rho = b$] is charged at the potential u_1 [resp., u_2].

2. Find the distribution of the potential inside a spherical capacitor if the sphere $\rho = a$ [resp., $\rho = b$] is maintained at the potential u_1 [resp., u_2].

3. One side of a right-angle parallelepiped is subject to a potential V, while the remaining sides are grounded. Find the distribution of the potential inside the parallelepiped.

4. An infinite $(-\infty < z < \infty)$ conducting cylinder is charged at the potential

$$V = \begin{cases} 1, & \text{if } 0 < \varphi < \pi, \\ 0, & \text{if } \pi < \varphi < 2\pi. \end{cases}$$

Find the distribution of the potential inside the cylindrical cavity.

5. Find the temperature distribution in an infinitely long $(-\infty < z < \infty)$ circular cylinder if the a heat flux $Q = q \cos \varphi$ per unit of length is given on its surface.

6. A constant current J passes through an infinite $(-\infty < z < \infty)$ coaxial cyclindrical cable $(a < \rho < b)$. Find the temperature distribution inside the cable if its inner surface $\rho = a$ is kept at temperature zero and the outer surface is thermally insulated.

7. Find the distribution of the potential in a thin plate shaped as a half-disc when the diameter of the half-disc is charged at potential V_1, while the remaining part of the boundary is charged at potential V_2.

8. Find the temperature distribution inside a thin rectangular plate if a constant heat flux Q is introduced through one of its sides, whereas the other three sides are kept at temperature zero.

9. Find the temperature distribution inside an infinite $(-\infty < z < \infty)$ circular cylinder if its surface is mantained at the temperature $A \cos \varphi + B \sin \varphi$, where A and B are constants.

10. Find the distribution of the potential inside an empty cylinder of radius R and height h whose two bases are grounded, whereas the lateral surface has the potential V.

11. Determine the steady temperature distribution inside a circular cylinder of finite length is a constant heat flux q is introduced through the lower base $z = 0$, whereas the lateral surface $\rho = a$ and the upper base are maintained at temperature zero.

12. Find the steady temperature distribution inside a homogeneous and isotropic ball if its surface is maintained at the temperature $A\sin^2\theta$ ($A=$ const).

13. Find the distribution of the potential in a spherical capacitor $1 < r < 2$ if the inner and outer plates have the potential $V_1 = \cos^2\theta$ and $V_2 = \frac{1}{8}(\cos^2\theta + 1)$, respectively.

14. Find the temperature distribution inside a spherical layer $1 < r < 2$ if the inner sphere is maintained at the temperature $T_1 = \sin\theta\sin\varphi$, whereas the outer sphere is maintained at the temperature of melting ice.

15. Solve the Dirichlet problem for the Poisson equation $\Delta u = e^y\sin x$ in the square $0 \le x \le \pi$, $0 \le y \le \pi$, with null boundary condition.

16. Solve the Dirichlet problem for the Poisson equation $\Delta u = x^4 - y^4$ in the disc of radius one, with null boundary condition.

17. Solve the Dirichlet problem for the Poisson equation $\Delta u = z$ in the ball of radius one, with null boundary condition.

18. Solve the Dirichlet problem for the Poisson equation $\Delta u = J_0\left(\frac{\mu_1}{R}\rho\right)$ in a cylinder of radius R and height h, with null boundary conditions.

19. Find the eigenoscillations of a rectangular membrane when two opposite edges are clamped and the other two are free.

20. Find the eigenoscillations of a circular cylinder under null boundary conditions of the first kind.

21. Find the steady concentration distribution of an unstable gas inside a sphere of radius a if a constant concentration u_0 is maintained at the surface of the sphere.

22. Solve the Dirichlet problem for the equation $\Delta u + k^2 u = 0$ in the interior and in the exterior of the sphere $\rho = R$ under the condition $u|_{\rho=R} = A$ ($A=$ const).

23. Solve the Neumann problem for the equation $\Delta u - k^2 u = 0$ in the interior and in the exterior of the sphere $\rho = R$ under the condition $\partial u/\partial n|_{\rho=R} = A$ ($A=$ const).

24. Find the steady distribution of potential in the first quadrant $x > 0$, $y > 0$ if the half-line $y = 0$ is grounded while the half-line $x = 0$ is maintained at the potential V.

25. Find the steady temperature distribution in the strip $0 < y < \pi$ if the temperature on the lower boundary $y = 0$ equals $A\cos x$ while the upper boundary is kept at the temperature of melting ice ($A=$ const).

26. Find the steady distribution of potential in the strip $0 < y < \pi$, $x > 0$ if the horizontal sides of the strip are grounded and the vertical side has the potential V.

27. Find the distribution of potential in an infinitely long eccentric cylindrical capacitor if the inner plate $|z + 1| = 9$ has the potential 1 while the outer plate $|z + 6| = 16$ is grounded.

28. Find the solution of the Dirichlet problem $\Delta u = 0$ in the domain $\operatorname{Im} z < 0$, $|z + 5i| > 3$ if

$$u|_{\operatorname{Im} z=0} = 0, \qquad u|_{|z+5i|=3} = 1.$$

29. Find the temperature distribution in the lower half-plane $y < 0$ if its boundary $y = 0$ is maintained at the temperature $A \sin x$ $(A = \text{const})$.

30. Find the temperature distribution in the upper half-plane $y > 0$ if its boundary $y = 0$ is maintained at the temperature $\theta(-x)$, where $\theta(x)$ is Heaviside function.

31. Find the distribution of potential in the upper half-space $z > 0$ if its boundary $z = 0$ has the potential $(1 + x^2 + y^2)^{-3/2}$.

32. Solve the Dirichlet problem for the Poisson equation $\Delta u = -e^{-z} \sin x \cos y$ in the half-space $z > 0$ with null boundary condition.

33. Find the steady temperature distribution in the exterior of a bounded circular cylinder ($\rho > 1$, $-\infty < z < \infty$) if the lateral surface ($\rho = 1$) is maintained at the temperature $u|_{\rho=1} = A \cos(2\varphi) + B \cos(5\varphi) + C \cos(10\varphi)$, where A, B, C are constants.

34. Solve the Dirichlet problem

$$\begin{cases} \Delta u = 0, & 0 < \rho < 1, \quad 0 \le \varphi \le 2\pi, \\ u|_{\rho=1} = \dfrac{\sin \varphi}{5 + 4 \cos \varphi} & 0 \le \varphi \le 2\pi. \end{cases}$$

35. Solve the following Neumann problem for the Laplace equation in the spherical layer $1 < \rho < 2$:

$$\begin{cases} \Delta u = 0 & \text{inside the layer,} \\ \dfrac{\partial u}{\partial n}\bigg|_{\rho=1} = P_2(\cos \theta), \\ \dfrac{\partial u}{\partial n}\bigg|_{\rho=2} = P_3(\cos \theta). \end{cases}$$

36. Solve the following boundary value problem in the disc $\{0 \le \rho \le a, 0 \le \varphi < 2\pi\}$:

$$\begin{cases} \Delta^2 u = x^2 + y^2 & \text{in the disc,} \\ u|_{\rho=a} = 0, \quad \dfrac{\partial u}{\partial n}\bigg|_{\rho=a} = 0. \end{cases}$$

37. Solve the following boundary value problem in the disc $\{0 \leq \rho \leq a, 0 \leq \varphi < 2\pi\}$:

$$
\begin{cases}
\Delta^2 u = 0 & \text{in the disc,} \\
u|_{\rho=a} = 1, \quad \dfrac{\partial u}{\partial n}\bigg|_{\rho=a} = \sin^3 \varphi.
\end{cases}
$$

38. Solve the following boundary value problem in the ball $\{0 \leq \rho \leq a\}$:

$$
\begin{cases}
\Delta^2 u = x^2 + y^2 + z^2 & \text{inside the ball,} \\
u|_{\rho=a} = 0, \quad \dfrac{\partial u}{\partial n}\bigg|_{\rho=a} = 0.
\end{cases}
$$

39. Solve the following boundary value problem in the half-space:

$$
\begin{cases}
\Delta^2 u = e^{-z} \sin x \cos y & -\infty < x, y < \infty, \quad z > 0, \\
u|_{z=0} = 0, \quad \dfrac{\partial u}{\partial z}\bigg|_{z=0} = 0.
\end{cases}
$$

40. Solve the following boundary value problem in the lower half-plane $(y < 0)$:

$$
\begin{cases}
\Delta u = 0 & -\infty < x < \infty, \quad y < 0, \\
u|_{y=0} = \dfrac{2x}{(1+x^2)^2}.
\end{cases}
$$

1.19. Answers

1. $u_1 = u_2 + (u_1 - u_2)\dfrac{\ln b/\rho}{\ln b/a}$.

2. $u = u_2 + (u_1 - u_2)\dfrac{1/r - 1/b}{1/a - 1/b}$.

3. $u = \displaystyle\sum_{n=1}^{\infty}\sum_{m=1}^{\infty} A_{nm} \sin\left(\dfrac{m\pi}{a}x\right) \sin\left(\dfrac{n\pi}{b}y\right) \sinh\left(\pi\sqrt{\dfrac{n^2}{a^2} + \dfrac{m^2}{b^2}}\, z\right)$, where

$$
A_{nm} = \begin{cases}
\dfrac{16V}{\pi^2 nm \sinh\left(\pi\sqrt{\dfrac{n^2}{a^2} + \dfrac{m^2}{b^2}}\, c\right)}, & \text{if } n \text{ and } m \text{ are odd,} \\[4mm]
0, & \text{if } n \text{ or } m \text{ is even,}
\end{cases}
$$

and a, b, c are the sides of the parallelepiped.

4. $u = \dfrac{1}{2} + \dfrac{1}{\pi}\arctan\dfrac{2a\rho\sin\varphi}{a^2 - \rho^2}$, where a is the radius of the cylinder.

5. $u = -\dfrac{q}{k}\rho\cos\varphi + c$, where k is the heat conduction coefficient of the cylinder.

6. $u = -\dfrac{q}{4}(\rho^2 - a^2) - \dfrac{qb^2}{2}\ln\dfrac{\rho}{a}$, where $q = -q_0/k$, $q_0 = 0.24J^2R$, R is the resistance per unit of length of the conductor, and k is the heat conduction coefficient.

7. $u = V_1 + \dfrac{2}{\pi}(V_2 - V_1)\arctan\dfrac{a\rho\sin\varphi}{\rho^2 - a^2}$.

8. $u = -\dfrac{4qa}{k\pi^2}\displaystyle\sum_{m=0}^{\infty}\dfrac{\sin\left[\frac{(2m+1)\pi}{a}x\right]}{(2m+1)^2}\cdot\dfrac{\sinh\left[\frac{(2m+1)\pi}{a}y\right]}{\cosh\left[\frac{(2m+1)\pi}{a}b\right]}$.

9. $u = A\dfrac{\rho}{a}\cos\varphi + B\dfrac{\rho}{a}\sin\varphi$, where a is the radius of the cylinder.

10. $u = \dfrac{4V}{\pi}\displaystyle\sum_{n=0}^{\infty}\dfrac{\sin\left[\frac{(2n+1)\pi}{h}z\right]}{2n+1}\cdot\dfrac{I_0\left(\frac{(2n+1)\pi}{h}\rho\right)}{I_0\left(\frac{(2n+1)\pi}{h}R\right)}$.

11. $u = \displaystyle\sum_{m=0}^{\infty}A_m\dfrac{\sinh\left[\frac{\mu_m}{a}(l-z)\right]}{\cosh\left[\frac{\mu_m}{l}z\right]}\cdot J_0\left(\dfrac{\mu_m}{a}\rho\right)$, where $A_m = \dfrac{2aq}{k\mu_m^2 J_1(\mu_m)}$, k is the heat conduction coefficient, and μ_m is the mth positive root of the equation $J_0(x) = 0$.

12. $u = \dfrac{2}{3}A - A\left(\dfrac{r^2}{a}\right)^2 \cdot \dfrac{3\cos^2\theta - 1}{3}$, where a is the radius of the ball.

13. $u = \dfrac{1}{3r} + \dfrac{3\cos^2\theta - 1}{3r^3}$.

14. $u = \dfrac{1}{7}\left(-r + \dfrac{8}{r^2}\right)\sin\theta\sin\varphi$.

15. $u = \dfrac{1}{2\sinh\pi}(ye^y\sinh\pi - \pi e^\pi\sinh y)\sin x$.

16. $u = \dfrac{1}{32}\rho^2(\rho^4 - 1)\cos(2\varphi)$.

17. $u = \dfrac{1}{10}(r^3 - r)\cos\theta.$

18. $u = \left\{ \dfrac{R^2}{\mu_1^2}\left[\cosh\left(\dfrac{\mu_1}{R}z\right) - 1\right] + \dfrac{R^2}{\mu_1^2}\left[1 - \cosh\left(\dfrac{\mu_1}{R}h\right)\right] \dfrac{\sin\left(\frac{\mu_1}{R}z\right)}{\sin\left(\frac{\mu_1}{R}h\right)} \right\} J_0\left(\dfrac{\mu_1}{R}\rho\right).$

19. $\lambda_{m,n} = \pi^2\left(\dfrac{m^2}{a^2} + \dfrac{n^2}{b^2}\right),\ m = 1,2\ldots,\ n = 1,2,\ldots,$ where a and b are the side

lengths of the membrane; $u_{m,n} = \sin\left(\dfrac{m\pi}{a}x\right)\cos\left(\dfrac{n\pi}{a}y\right).$

20. $\lambda_{m,n,k} = \left(\dfrac{k\pi}{h}\right)^2 + \left(\dfrac{\mu_m^{(n)}}{a}\right)^2,\ n = 0,1,\ldots,\ m,k = 1,2,\ldots,$ where $\mu_m^{(n)}$ is the

mth positive root of the equation $J_n(x) = 0$, h is the height of the cylinder, and a is its radius;

$$v_{n,m,k} = \sin\left(\dfrac{k\pi}{h}z\right) J_n\left(\dfrac{\mu_m^{(n)}}{a}\rho\right) \left\{ \begin{array}{c} \cos(n\varphi) \\ \sin(n\varphi) \end{array} \right. .$$

21. $u = u_0\dfrac{a}{\rho}\cdot\dfrac{\sinh(k\rho)}{\sinh(ka)},$ where k is taken from the equation $\Delta u - k^2 u = 0$.

22. $u = \dfrac{AR}{\rho}\cdot\dfrac{\sin(k\rho)}{\sin(kR)}$ if $\rho \le R$, and $u = \dfrac{AR}{\rho}\cdot\dfrac{e^{ik\rho}}{e^{ikR}}$ if $\rho \ge R.$

23. $u = \dfrac{AR^2\sinh(k\rho)}{\rho[kR\cosh(kR) - \sinh(kR)]}$ if $\rho \le R$, and $u = -\dfrac{aR^2}{\rho}\cdot\dfrac{e^{k(R-\rho)}}{1+kR}$ if $\rho \ge R.$

24. $u = \dfrac{2V}{\pi}\arctan\dfrac{y}{x}.$

25. $u = \dfrac{A}{\sinh\pi}\cos x\,\sinh(\pi - y).$

26. $u = \dfrac{V}{\pi}\arctan\left(\dfrac{2\sinh x\,\sin y}{\sinh^2 x - \sin^2 y}\right).$

27. $u = \dfrac{1}{\ln(2/3)}\left(\ln 2 + \ln\left|\dfrac{z-2}{z-26}\right|\right).$

28. $u = \dfrac{1}{\ln 3}\ln\left|\dfrac{z+4i}{z-4i}\right|.$

29. $u = Ae^y \sin x.$

30. $u = \dfrac{1}{2} - \dfrac{1}{\pi} \arctan \dfrac{x}{y}.$

31. $u = \dfrac{z+1}{[x^2 + (z+1)^2 + y^2]^{3/2}}.$

32. $u = (e^{-\sqrt{2}z} - e^{-z}) \sin x \cos y.$

33. $u = \dfrac{A}{\rho^2} \cos(2\varphi) + \dfrac{B}{\rho^5} \cos(5\varphi) + \dfrac{C}{\rho^{10}} \cos(10\varphi).$

34. $u = \dfrac{\rho \sin \varphi}{\rho^2 + 4\rho \cos \varphi + 4}.$

35. $u = \dfrac{1}{31} \left(\dfrac{\rho^2}{2} + \dfrac{32}{2} \cdot \dfrac{1}{\rho^3} \right) P_2(\cos \theta) + \dfrac{1}{47} \left(4\rho^3 + \dfrac{3}{\rho^4} \right) P_3(\cos \theta) + C,$
where C is an arbitrary constant.

36. $u = \dfrac{a^6}{576} \left[\left(\dfrac{\rho}{a} \right)^6 - 3 \left(\dfrac{\rho}{a} \right)^2 + 2 \right].$

37. $u = 1 - \dfrac{a^2 - \rho^2}{2a} \left[\dfrac{3}{4} \left(\dfrac{\rho}{a} \right) \sin \varphi - \dfrac{1}{4} \left(\dfrac{\rho}{a} \right)^3 \sin(3\varphi) \right].$

38. $u = \dfrac{a^6}{840} \left[\left(\dfrac{\rho}{a} \right)^6 - 3 \left(\dfrac{\rho}{a} \right)^2 + 2 \right].$

39. $u = \left[Ae^{-\sqrt{2+\sqrt{2}}z} + (1 - A)e^{-\sqrt{2-\sqrt{2}}z} - e^{-z} \right] \sin x \cos y,$ where

$$A = \dfrac{1}{2} \sqrt{1 + \dfrac{1}{\sqrt{2}}} + \dfrac{1}{2} \sqrt{1 - \dfrac{1}{\sqrt{2}}} - \dfrac{1}{\sqrt{2}}.$$

40. $u = \dfrac{2x(1-y)}{[x^2 + (1-y)^2]^2}.$

Chapter 2
Hyperbolic problems

Up to this point we have considered steady physical processes, which were described by elliptic equations. We now turn our attention to the study of another class of partial differential equations – the hyperbolic equations. We begin with the one-dimensional wave equation $u_{tt} = a^2 u_{xx}$ which describes, among other processes, the transversal oscillations of a string.

2.1. The travelling-wave method

The one-dimensional case (one space variable). Let us consider the *Cauchy problem*

$$\begin{cases} u_{tt} = a^2 u_{xx}, & -\infty < x < \infty, \quad t > 0, \\ u(x,0) = f(x), & u_t(x,0) = g(x), \quad -\infty < x < \infty, \end{cases}$$

i.e., the function $u(x,t)$ is sought in the upper half-plane $t > 0$ of the (x,t)-plane (Figure 2.1).

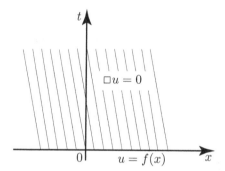

FIGURE 2.1. $\Box u = \dfrac{\partial^2 u}{\partial t^2} - a^2 \dfrac{\partial^2 u}{\partial x^2}$

It is well known that this problem describes the motion of an infinite string with given initial conditions. Its solution is provided by the *d'Alembert formula*

$$u(x,t) = \frac{f(x-at) + f(x+at)}{2} + \frac{1}{2a} \int_{x-at}^{x+at} g(\zeta)\, d\zeta.$$

This formula can be rewritten as

$$u(x,t) = \frac{f(x-at) + f(x+at)}{2} + \frac{1}{2a}[G(x+at) - G(x-at)],$$

where $G'(\zeta) = g(\zeta)$.

From the physical point of view this solution is interesting because it represents a sum of two travelling waves, which move in opposite directions with velocity a (Figure 2.2).

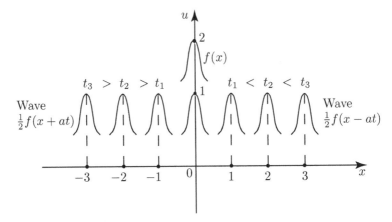

FIGURE 2.2.

Let us give a space-time interpretation of d'Alembert's formula. In the (x,t)-plane (Figure 2.3), let us take an arbitrary point (x_0, t_0) and draw through it two lines: $x - at = x_0 - at_0$ and $x + at = x_0 + at_0$ (these lines are called *characteristics of the wave equation* $u_{tt} = a^2 u_{xx}$). Then the (value of the) solution $u(x, t)$ at the point (x_0, t_0) can be interpreted as the mean value of the function $f(x)$ at the points $x_1 = x_0 - at_0$ and $x_2 = x_0 + at_0$ plus the integral from x_1 to x_2, multiplied by $1/(2a)$, of the initial speed $g(x)$.

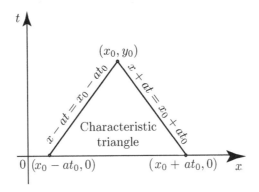

FIGURE 2.3.

Remark. If we consider the nonhomogeneous problem with homogeneous (null) initial conditions

$$\begin{cases} u_{tt} = a^2 u_{xx} + f(x,t), & -\infty < x < \infty, \quad t > 0, \\ u(x,0) = 0, \quad u_t(x,0) = 0, & -\infty < x < \infty, \end{cases}$$

then using the *Duhamel's method* or *principle* (or the *method of impulses* one can obtain the solution in the form

$$u(x,t) = \frac{1}{2a} \int_0^t d\tau \int_{x-a(t-\tau)}^{x+a(t-\tau)} f(\zeta,\tau)\,d\zeta.$$

Example 1 [4, Ch. II, no. 57]. Suppose that a wave of the form $\varphi(x - at)$ travels along an infinite string. Taking this wave as the initial condition at time $t = 0$, find the state of the string at time $t > 0$.

Solution. In our problem the travelling wave at time $t = 0$ characterized by nonzero "initial" deviations and velocities

$$u(x,0) = \varphi(x), \qquad u_t(x,0) = -a\varphi'(x), \qquad -\infty < x < \infty.$$

Therefore, we obtain the Cauchy problem

$$\begin{cases} u_{tt} = a^2 u_{xx}, & -\infty < x < \infty, \quad t > 0, \\ u(x,0) = \varphi(x), \quad u_t(x,0) = -a\varphi'(x), & -\infty < x < \infty. \end{cases}$$

Using d'Alembert's formula we get

$$u(x,t) = \frac{\varphi(x-at) + \varphi(x+at)}{2} + \frac{1}{2a} \int_{x-at}^{x+at} [-a\varphi'(\zeta)]d\zeta =$$

$$= \frac{\varphi(x-at) + \varphi(x+at)}{2} + \frac{1}{2a}[-a\varphi(x+at) + a\varphi(x-at)] = \varphi(x-at).$$

Example 2 [6, Ch. II, 21.23]. Find the state of the loaded semi-infinite string in the problem

$$\begin{cases} u_{tt} = u_{xx} - 6, & x > 0, \quad t > 0, \\ u(x,0) = x^2, \quad u_t(x,0) = 0, & x \geq 0, \\ u_t(0,t) + 2u_x(0,t) = -4t, & t \geq 0, \end{cases} \tag{2.1}$$

i.e., the function $u(x,t)$ is sought in the first quadrant $x > 0$, $t > 0$ of the (x,t)-plane (Figure 2.4).

Solution. First let us find a particular solution of the equation $u_{tt} = u_{xx} - 6$, without paying attention to the initial and boundary conditions. We shall assume that this particular solution u_1 depends only on x, i.e., $u_1 = u_1(x)$. Then in order

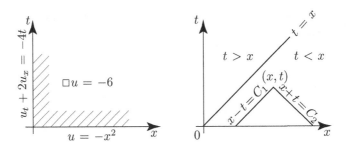

FIGURE 2.4.   FIGURE 2.5

to find it we must solve the equation $d^2u_1/dx^2 = 6$, which yields $u_1(x) = 3x^2$. Now let us consider the new function $v(x,t) = u(x,t) - u_1(x)$. Then from equation (2.1) we obtain the equation $v_{tt} = v_{xx}$ and the problem from which one determines $v(x,t)$:

$$\begin{cases} v_{tt} = v_{xx}, & x > 0, \quad t > 0, \\ v(x,0) = -2x^2, & v_t(x,0) = 0, \quad x \geq 0, \\ v_t(0,t) + 2v_x(0,t) = -4t, & t \geq 0, \end{cases} \qquad (2.2)$$

Clearly, for $t < x$ the solution $v(x,t)$ can be written via d'Alembert formula in the form (see Figure 2.5)

$$v(x,t) = -(x-t)^2 - (x+t)^2 = -2(x^2 + t^2).$$

For $t > x$ we will seek the solution of problem (2.2) in the form

$$v(x,t) = f(x+t) + g(t-x).$$

To find the functions $f(x)$ and $g(x)$ we use the continuity of the solution $v(x,t)$ on the characteristic $t = x$ and the boundary condition $v_t(0,t) + 2v_x(0,t) = -4t$. Indeed, on the characteristic $t = x$ we have $f(2t) + g(0) = -4t^2$, whence $f(t) = -t^2 - g(0)$. Further, from the boundary condition in (2.2) it follows that

$$g'(t) + f'(t) + 2f'(t) - 2g'(t) = -4t,$$

which after integration yields

$$3f(t) - g(t) = -2t^2 + 3f(0) - g(0).$$

Clearly, $f(0) + g(0) = 0$. Thus, we arrived at a system of two equations for the two unknown functions $f(t)$ and $g(t)$:

$$\begin{cases} f(t) = -t^2 + f(0), \\ 3f(t) - g(t) = -2t^2 + 4f(0). \end{cases}$$

This gives $g(t) = -t^2 - f(0)$. Thus, for $t > x$,

$$v(x,t) = -(t+x)^2 + f(0) - (t-x)^2 - f(0) = -2(x^2 + t^2).$$

We conclude that the sought solution is given for $t > 0$, $x > 0$ by

$$u(x,t) = v(x,t) + u_1(x) = x^2 - 2t^2.$$

Example 3 [4, Ch. II, no. 78]. Consider an infinite elastic beam constructed by joining in the point $x = 0$ two homogeneous semi-infinite beams. For $x < 0$ [resp., $x > 0$] the mass density, the elasticity modulus, and the speed of propagation of small transversal perturbations are equal to ρ_1, E_1, and a_1 [resp., ρ_2, E_2, and a_2]. Assume that from the domain $x < 0$ a wave $u_1(x,t) = f(t - x/a_1)$ travels along the beam. Study the behaviour of solution when $E_2 \to 0$ and $E_2 \to \infty$.

Solution. The deviations of the points of the beam are governed by the equations

$$\begin{cases} u_{1tt} = a_1^2 u_{1xx}, & -\infty < x < 0, \quad t > 0, \\ u_{2tt} = a_2^2 u_{2xx}, & 0 < x < \infty, \quad t > 0, \end{cases} \tag{2.3}$$

and the initial conditions

$$\begin{cases} u_1(x,0) = f(-x/a_1), & u_{1t}(x,0) = f'(-x/a_1), & -\infty < x < 0, \\ u_2(x,0) = 0, & u_{2t}(x,0) = 0, & 0 < x < \infty. \end{cases} \tag{2.4}$$

These equations and initial conditions must be supplemented by the conjugation conditions

$$\begin{cases} u_1(0,t) = u_2(0,t) & \text{continuity of displacements}, \\ E_1 u_{1x}(0,t) = E_2 u_{2x}(0,t) & \text{continuity of stresses}. \end{cases}$$

Let us seek the unknown functions in the form

$$u_1(x,t) = f_1\left(t - \frac{x}{a_1}\right) + g_1\left(t + \frac{x}{a_1}\right), \qquad -\infty < x < 0, \quad t > 0,$$

$$u_2(x,t) = f_2\left(t - \frac{x}{a_2}\right) + g_2\left(t + \frac{x}{a_2}\right), \qquad -0 < x < \infty, \quad t > 0.$$

The four functions $f_1(s)$, $g_1(s)$, $f_2(s)$, $g_2(s)$ are determined by the initial conditions and conjugation conditions. First of all, using (2.3) we can write

$$\begin{cases} f_1\left(-\dfrac{x}{a_1}\right) + g_1\left(\dfrac{x}{a_1}\right) = f\left(-\dfrac{x}{a_1}\right), \\ f_1'\left(-\dfrac{x}{a_1}\right) + g_1'\left(\dfrac{x}{a_1}\right) = f'\left(-\dfrac{x}{a_1}\right), \end{cases} \tag{2.5}$$

$$\begin{cases} f_2\left(-\dfrac{x}{a_2}\right) + g_2\left(\dfrac{x}{a_2}\right) = 0, \\[3mm] f_2'\left(-\dfrac{x}{a_2}\right) + g_2'\left(\dfrac{x}{a_2}\right) = 0. \end{cases} \tag{2.6}$$

Integrating the second equation in (2.5) and in (2.6) and denoting $x/a_1 = s$, $x/a_2 = z$, we obtain

$$\begin{cases} f_1(-s) + g_1(s) = f(-s), \\ -f_1(-s) + g_1(s) = -f(-s), \end{cases} \tag{2.7}$$

for $s < 0$, and

$$\begin{cases} f_2(-z) + g_2(z) = 0, \\ -f_2(-z) + g_2(z) = 0 \end{cases} \tag{2.8}$$

for $z > 0$. Solving the system (2.7) we find

$$f_1(-s) = f(-s), \qquad g_1(s) = 0, \qquad s < 0.$$

Hence,

$$u_1(x,t) = \begin{cases} f\left(t - \dfrac{x}{a_1}\right) + g_1\left(t + \dfrac{x}{a_1}\right), & \text{if } t + \dfrac{x}{a_1} > 0, \\[3mm] f\left(t - \dfrac{x}{a_1}\right), & \text{if } t + \dfrac{x}{a_1} < 0. \end{cases}$$

Next, solving the systems (2.8) we find

$$f_2(-z) = 0, \qquad g_2(z) = 0, \qquad z > 0,$$

whence

$$u_2(x,t) = \begin{cases} f_2\left(t - \dfrac{x}{a_2}\right), & \text{if } t - \dfrac{x}{a_2} > 0, \\[3mm] 0, & \text{if } t - \dfrac{x}{a_2} < 0. \end{cases}$$

Now let us use the conjugation conditions to find the functions $f_2(s)$ and $g_2(s)$. From the equality $u_1(0,t) = u_2(0,t)$ it follows that

$$f(t) + g_1(t) = f_2(t).$$

Similarly, from the equality $E_1 u_{1x}(0,t) = E_2 u_{2x}(0,t)$ it follows that

$$E_1\left[-\frac{1}{a_1}f_1'(t) + \frac{1}{a_1}g_1'(t)\right] = E_2\left[-\frac{1}{a_2}f_2'(t) + \frac{1}{a_2}g_2'(t)\right].$$

Since $a_1 = \sqrt{E_1/\rho_1}$ and $a_2 = \sqrt{E_2/\rho_2}$, we obtain

$$\sqrt{E_1\rho_1}\left[-f'(t) + g_1'(t)\right] = \sqrt{E_2\rho_2}\left[-f_2'(t)\right],$$

which after integration yields

$$\sqrt{E_1\rho_1}\left[-f(t) + g_1(t)\right] = \sqrt{E_2\rho_2}\left[-f_2(t)\right].$$

Therefore, we have the following system of equations for the desired functions $f_2(t)$ and $g_1(t)$:

$$\begin{cases} f(t) + g_1(t) = f_2(t), \\ \sqrt{E_1\rho_1}\left[-f(t) + g_1(t)\right] = \sqrt{E_2\rho_2}\left[-f_2(t)\right]. \end{cases}$$

Solving this system we get

$$g_1(t) = \frac{\sqrt{E_1\rho_1} - \sqrt{E_2\rho_2}}{\sqrt{E_1\rho_1} + \sqrt{E_2\rho_2}}\, f(t),$$

$$f_2(t) = \frac{2\sqrt{E_1\rho_1}}{\sqrt{E_1\rho_1} + \sqrt{E_2\rho_2}}\, f(t).$$

We conclude that the solution of our problem is represented by the following pair of functions:

$$u_1(x,t) = \begin{cases} f\left(t - \frac{x}{a_1}\right) + \dfrac{\sqrt{E_1\rho_1} - \sqrt{E_2\rho_2}}{\sqrt{E_1\rho_1} + \sqrt{E_2\rho_2}}\, f\left(t + \frac{x}{a_1}\right), & \text{if } t + \frac{x}{a_1} > 0, \\ f\left(t - \frac{x}{a_1}\right), & \text{if } t + \frac{x}{a_1} < 0, \end{cases}$$

$$u_2(x,t) = \begin{cases} \dfrac{2\sqrt{E_1\rho_1}}{\sqrt{E_1\rho_1} + \sqrt{E_2\rho_2}}\, f\left(t - \frac{x}{a_2}\right), & \text{if } t - \frac{x}{a_2} > 0, \\ f\left(t - \frac{x}{a_2}\right), & \text{if } t - \frac{x}{a_2} < 0. \end{cases}$$

The term

$$\frac{\sqrt{E_1\rho_1} - \sqrt{E_2\rho_2}}{\sqrt{E_1\rho_1} + \sqrt{E_2\rho_2}}\, f\left(t + \frac{x}{a_1}\right)$$

in $u_1(t,x)$ for $t + x/a_1 > 0$ is a reflected wave. Obviously, this wave is absent if $E_1\rho_1 = E_2\rho_2$. If $E_2 = 0$, then the reflected wave is $f(t + x/a_1)$; if $E_2 = \infty$, then the reflected wave is $-f(t + x/a_1)$.

The refracted wave is $u_2(x,t)$. If $E_2 = 0$, then the amplitude of this wave is twice that of the incident wave; if $E_2 = \infty$, then there is no refracted wave.

Example 4 [4, Ch. III, no. 442]. Solve the Cauchy problem

$$\begin{cases} u_{tt} = u_{xx} + x\sin t, & -\infty < x < \infty, \quad t > 0, \\ u(x,0) = \sin x, \quad u_t(x,0) = \cos x, & -\infty < x < \infty. \end{cases}$$

Solution. Let us apply the d'Alembert formula

$$u(x,t) = \frac{\varphi(x - at) + \varphi(x + at)}{2} +$$

$$+ \frac{1}{2a}\int_{x-at}^{x+at} \psi(\zeta)\, d\zeta + \frac{1}{2a}\int_0^t d\tau \int_{x-a(t-\tau)}^{x+a(t-\tau)} f(\zeta,\tau)\, d\zeta,$$

which gives the solution of the problem

$$\begin{cases} u_{tt} = a^2 u_{xx} + f(x,t), & -\infty < x < \infty, \quad t > 0, \\ u(x,0) = \varphi(x), & u_t(x,0) = \psi(x), \quad -\infty < x < \infty. \end{cases}$$

In our case

$$\varphi(x) = \sin x, \qquad \psi(x) = \cos x, \qquad f(x,t) = x \sin t, \qquad a = 1.$$

Therefore, the solution sought has the form

$$u(x,t) = \frac{\sin(x-t) + \sin(x+t)}{2} +$$

$$+ \frac{1}{2} \int_{x-t}^{x+t} \cos \zeta \, d\zeta + \frac{1}{2} \int_0^t d\tau \int_{x-(t-\tau)}^{x+(t-\tau)} \zeta \sin \tau \, d\zeta =$$

$$= \frac{\sin(x-t) + \sin(x+t)}{2} + \frac{\sin(x+t) - \sin(x-t)}{2} +$$

$$+ \frac{1}{2} \int_0^t \sin \tau \, d\tau \cdot \frac{\zeta^2}{2} \Big|_{x-(t-\tau)}^{x+(t-\tau)} =$$

$$= \sin(x+t) + x \int_0^t (t-\tau) \sin \tau \, d\tau = sin(x+t) + x(t - \sin t).$$

The two-dimensional case (two space variables). Here we will consider the two-dimensional wave equation. This equation described the propagation of sound in a gas, the oscillations of a membrane, the propagation of electromagnetic waves, and so on.

By definition, *the Cauchy problem for the two-dimensional wave equation* is the problem of finding a function $u(x,y,t)$ that satisfies the equation

$$u_{tt} = a^2 (u_{xx} + u_{yy}) + f(x,y,t), \qquad -\infty < x, y < \infty, \qquad t > 0,$$

and the initial conditions

$$u(x,y,0) = \varphi(x,y), \qquad u_t(x,y,0) = \psi(x,y), \qquad \infty < x, y < \infty,$$

i.e., the function $u(x,y,t)$ is sought in the upper half-space $t > 0$ (Figure 2.6).

The solution of this problem exists, is unique, and is given by the Poisson formula

$$u(x,y,t) = \frac{1}{2\pi a} \left[\frac{\partial}{\partial t} \iint_{K_{at}} \frac{\varphi(\zeta,\eta)}{\sqrt{a^2 t^2 - \rho^2}} \, d\zeta \, d\eta + \iint_{K_{at}} \frac{\psi(\zeta,\eta)}{\sqrt{a^2 t^2 - \rho^2}} \, d\zeta \, d\eta + \right.$$

$$\left. + \int_0^t d\tau \iint_{K_{a(t-\tau)}} \frac{f(\zeta,\eta,\tau)}{\sqrt{a^2 (t-\tau)^2 - \rho^2}} \, d\zeta \, d\eta \right],$$

where $\rho^2 = (\zeta - x)^2 + (\eta - y)^2$ and K_{as} denotes the disc of radius as centered at the point (x,y) (Figure 2.7).

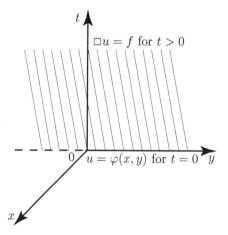

FIGURE 2.6.

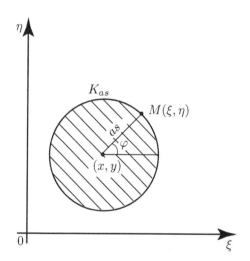

FIGURE 2.7.

Example 5 [6, Ch. IV, 12, 37(2)]. Solve the Cauchy problem

$$\begin{cases} u_{tt} = u_{xx} + u_{yy} + 6xyt, & -\infty < x, y < \infty, & t > 0, \\ u(x,y,0) = x^2 - y^2, & u_t(x,y,0) = xy, & -\infty < x, y < \infty. \end{cases}$$

Solution. Let us apply the Poisson formula, setting

$$\varphi(x,y) = x^2 - y^2, \qquad \psi(x,y) = xy, \qquad f(x,y,t) = 6xyt, \qquad a = 1.$$

We have

$$u(x,y,t) = \frac{1}{2\pi} \frac{\partial}{\partial t} \iint_{K_t} \frac{\zeta^2 - \eta^2}{\sqrt{t^2 - (\zeta - x)^2 - (\eta - y)^2}} \, d\zeta \, d\eta +$$

$$+ \frac{1}{2\pi} \iint_{K_t} \frac{\zeta\eta}{\sqrt{t^2 - (\zeta - x)^2 - (\eta - y)^2}} \, d\zeta \, d\eta +$$

$$+ \frac{1}{2\pi} \int_0^t d\tau \iint_{K_{t-\tau}} \frac{6\zeta\eta\tau}{\sqrt{(t - \tau)^2 - (\zeta - x)^2 - (\eta - y)^2}} \, d\zeta \, d\eta.$$

The integration is carried out over the disks K_t and $K_{t-\tau}$. We can write

$$\begin{cases} \zeta = x + \rho\cos\varphi, \\ \eta = y + \rho\sin\varphi, \end{cases} \quad \text{where } 0 \le \rho \le s = \begin{cases} t, \\ t - \tau, \end{cases} \quad 0 \le \varphi \le 2\pi$$

Let us calculate separately each of the integrals written above:

$$J_1 = \frac{1}{2\pi} \frac{\partial}{\partial t} \int_0^{2\pi} \int_0^t \frac{(x + \rho\cos\varphi)^2 - (y + \rho\sin\varphi)^2}{\sqrt{t^2 - \rho^2}} \rho \, d\rho \, d\varphi =$$

$$= \frac{1}{2\pi} \frac{\partial}{\partial t} \int_0^{2\pi} \int_0^t \frac{x^2 - y^2}{\sqrt{t^2 - \rho^2}} \rho \, d\rho \, d\varphi + \frac{1}{2\pi} \frac{\partial}{\partial t} \int_0^t \frac{2x\rho^2 \, d\rho}{\sqrt{t^2 - \rho^2}} \int_0^{2\pi} \cos\varphi \, d\varphi -$$

$$- \frac{1}{2\pi} \frac{\partial}{\partial t} \int_0^t \frac{2y\rho^2 \, d\rho}{\sqrt{t^2 - \rho^2}} \int_0^{2\pi} \cos\varphi \, d\varphi + \frac{1}{2\pi} \frac{\partial}{\partial t} \int_0^t \frac{\rho^2 \, d\rho}{\sqrt{t^2 - \rho^2}} \int_0^{2\pi} \cos(2\varphi) \, d\varphi =$$

$$= \frac{x^2 - y^2}{2\pi} \frac{\partial}{\partial t} \left(2\pi \int_0^t \frac{\rho^2 \, d\rho}{\sqrt{t^2 - \rho^2}} \right) = x^2 - y^2;$$

$$J_2 = \frac{1}{2\pi} \int_0^{2\pi} \int_0^t \frac{(x + \rho\cos\varphi)(y + \rho\sin\varphi)}{\sqrt{t^2 - \rho^2}} \rho \, d\rho \, d\varphi =$$

$$= \frac{1}{2\pi} \int_0^t \frac{xy\rho}{\sqrt{t^2 - \rho^2}} \, d\rho \int_0^{2\pi} d\varphi + \frac{1}{2\pi} \int_0^t \frac{x\rho^2}{\sqrt{t^2 - \rho^2}} \, d\rho \int_0^{2\pi} \sin\varphi \, d\varphi +$$

$$+ \frac{1}{2\pi} \int_0^t \frac{y\rho^2}{\sqrt{t^2 - \rho^2}} \, d\rho \int_0^{2\pi} \cos\varphi \, d\varphi +$$

$$+ \frac{1}{2\pi} \int_0^t \frac{\rho^2}{\sqrt{t^2 - \rho^2}} \, d\rho \int_0^{2\pi} \sin\varphi\cos\varphi \, d\varphi = xyt.$$

Finally,

$$J_2 = \frac{1}{2\pi} \int_0^t d\tau \int_0^{2\pi} \int_0^{t-\tau} \frac{6\tau(x + \rho\cos\varphi)(y + \rho\sin\varphi)}{\sqrt{(t - \tau)^2 - \rho^2}} \rho \, d\rho \, d\varphi =$$

$$= \frac{3}{\pi} \int_0^t \tau \, d\tau \int_0^{t-\tau} \frac{xy\rho}{\sqrt{(t-\tau)^2 - \rho^2}} \, d\rho \int_0^{2\pi} d\varphi =$$

$$= 6xy \int_0^t \tau \, d\tau \int_0^{t-\tau} \frac{\rho}{\sqrt{(t-\tau)^2 - \rho^2}} \, d\rho = 6xy \int_0^t (t-\tau)\tau \, d\tau = xyt^3.$$

Therefore, the sought solution is

$$u(x, y, t) = J_1 + J_2 + J_3 = xyt(1 + t^2) + x^2 - y^2.$$

The three-dimensional case (three space variables). By definition, the *Cauchy problem for the three-dimensional wave equation* is the problem of finding a function $u(x, y, z, t)$ that satisfies the equation

$$u_{tt} = a^2 \Delta u + f(x, y, z, t), \qquad -\infty < x, y, z < \infty, \qquad t > 0,$$

and the initial conditions

$$u(x, y, z, 0) = \varphi(x, y, z), \qquad u_t(x, y, z, 0) = \psi(x, y, z), \qquad -\infty < x, y, z < \infty.$$

The solution of this problem exists, is unique, and is given by Kirchhoff's formula

$$u(x, y, z, t) = \frac{1}{4\pi a^2} \left[\frac{\partial}{\partial t} \iint_{S_{at}} \frac{\varphi(\xi, \eta, \zeta)}{t} \, d\sigma + \iint_{S_{at}} \frac{\psi(\xi, \eta, \zeta)}{t} \, d\sigma + \right.$$

$$\left. + \int_0^t d\tau \iint_{S_{a(t-\tau)}} \frac{f(\xi, \eta, \zeta, \tau)}{t - \tau} \, d\sigma \right],$$

where S_{ap} denotes the sphere of radius ap centered at the point (x, y, z) (Figure 2.8).

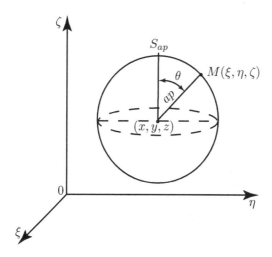

FIGURE 2.8.

Example 6 [6, Ch. IV, 12.38(1)]. Solve the Cauchy problem

$$\begin{cases} u_{tt} = u_{xx} + u_{yy} + u_{zz} + 2xyz, & -\infty < x, y, z < \infty, \quad t > 0, \\ u(x,y,z,0) = x^2 + y^2 - 2z^2, & u_t(x,y,z,0) = 1, \quad -\infty < x, y, z < \infty. \end{cases}$$

Solution. Let us apply the Kirchhoff formula, setting

$$\varphi(x,y,z) = x^2 + y^2 - 2z^2, \quad \psi(x,y,z) = 1, \quad f(x,y,z,t) = 2xyz, \quad a = 1.$$

We obtain

$$u(x,y,z,t) = \frac{1}{4\pi} \frac{\partial}{\partial t} \iint_{S_t} \frac{\xi^2 + \eta^2 - 2\zeta^2}{t} \, d\sigma +$$

$$+ \frac{1}{4\pi} \iint_{S_t} \frac{1}{t} \, d\sigma + \frac{1}{4\pi} \int_0^t d\tau \iint_{S_{t-\tau}} \frac{2\xi\eta\zeta}{t-\tau} \, d\sigma.$$

Here the integration is carried out over the spheres S_t and $S_{t-\tau}$ of radii t and $t-\tau$, respectively, centered at the point (x, y, z) (see Figure 2.8). If $M(\xi, \eta, \zeta) \in S_t$, then clearly we can write

$$\xi = x + t \sin\theta \cos\varphi,$$
$$\eta = y + t \sin\theta \sin\varphi,$$
$$\zeta = z + t \cos\theta,$$

where $0 \le \varphi \le 2\pi$, $0 \le \theta \le \pi$. Let us calculate separately each of the above integrals:

$$J_1 = \frac{1}{4\pi} \frac{\partial}{\partial t} \int_0^{2\pi} \int_0^\pi \left[\frac{(x + t \sin\theta \cos\varphi)^2 + (y + t \sin\theta \sin\varphi)^2}{t} - \right.$$

$$\left. - \frac{2(x + t \cos\theta)^2}{t} \right] t^2 \sin\theta \, d\theta \, d\varphi =$$

$$= \frac{1}{4\pi} \frac{\partial}{\partial t} \int_0^{2\pi} \int_0^\pi \left(x^2 + 2tx \sin\theta \cos\varphi + t^2 \sin^2\theta \cos^2\varphi \right) t^2 \sin\theta \, d\theta \, d\varphi +$$

$$+ \frac{1}{4\pi} \frac{\partial}{\partial t} \int_0^{2\pi} \int_0^\pi \left(y^2 + 2ty \sin\theta \sin\varphi + t^2 \sin^2\theta \sin^2\varphi \right) t^2 \sin\theta \, d\theta \, d\varphi +$$

$$+ \frac{1}{4\pi} \frac{\partial}{\partial t} \int_0^{2\pi} \int_0^\pi \left(-2z^2 - 4tz \cos\theta - 2t^2 \cos^2\theta \right) t^2 \sin\theta \, d\theta d\varphi =$$

$$= \frac{1}{4\pi} \frac{\partial}{\partial t} \int_0^{2\pi} \int_0^\pi t^2 \left(x^2 + y^2 - 2z^2 \right) \sin\theta \, d\theta \, d\varphi +$$

$$+ \frac{1}{4\pi} \frac{\partial}{\partial t} \int_0^{2\pi} \int_0^\pi t^3 \left(\sin^2\theta - 2\cos^2\theta \right) \sin\theta \, d\theta \, d\varphi +$$

$$+ \frac{1}{4\pi} \frac{\partial}{\partial t} \int_0^\pi 2t^2 x \sin^2 \theta \, d\theta \int_0^{2\pi} \cos \varphi \, d\varphi +$$

$$+ \frac{1}{4\pi} \frac{\partial}{\partial t} \int_0^\pi 2t^2 y \sin^2 \theta \, d\theta \int_0^{2\pi} \sin \varphi \, d\varphi -$$

$$- \frac{1}{4\pi} \frac{\partial}{\partial t} \int_0^\pi 2t^2 z \sin \theta \cos \theta \, d\theta \int_0^{2\pi} d\varphi = x^2 + y^2 - 2z^2,$$

$$J_2 = \frac{1}{4\pi} \frac{\partial}{\partial t} \int_0^{2\pi} \int_0^\pi \frac{1}{t} t^2 \sin \theta \cos \theta \, d\theta \, d\varphi = t,$$

and

$$J_3 = \frac{1}{4\pi} \int_0^t d\tau \int_0^{2\pi} \int_0^\pi \frac{2(x + (t-\tau) \sin \theta \cos \varphi)(y + (t-\tau) \sin \theta \sin \varphi)}{t - \tau} \times$$

$$\times (z + (t-\tau) \cos \theta)(t-\tau)^2 \sin \theta \, d\theta \, d\varphi =$$

$$= \frac{1}{2\pi} \int_0^t (t-\tau) d\tau \int_0^{2\pi} \int_0^\pi xyz \sin \theta \, d\theta \, d\varphi = 2xyz \int_0^t (t-\tau) d\tau = t^2 xyz.$$

Therefore, the solution to our Cauchy problem is

$$u(x, y, z, t) = x^2 + y^2 - 2z^2 + t + t^2 xyz.$$

2.2. The method of selection of particular solutions

Sometimes it is convenient, instead of using the general formula that gives the solution of the Cauchy problem for the wave equation, to take advantage of the specific form of the right-hand side of the equation and of the initial conditions. Indeed, the d'Alembertian $\Box u = \partial^2 u/\partial t^2 - \Delta u$, defined on twice-differentiable functions, takes, for example, a function of the form $u(x, y, z, t) = e^{ax} \sin(by) \cos(cz) P_m(t)$, where $P_m(t)$ is a polynomial of degree m, into a function of the same form. Using this observation, it is easy to choose the requisite particular solution of the Cauchy problem.

Example 1. Solve the Cauchy problem

$$\begin{cases} u_{tt} = \Delta u + te^{5x} \sin(3y) \cos(4z), & -\infty < x, y, z < \infty, \quad t > 0, \\ u|_{t=0} = e^{6x+8y} \cos(10z), \quad u_t|_{t=0} = e^{3y+4z} \sin(5x), & -\infty < x, y, z < \infty. \end{cases}$$

Solution Let us decompose the problem into two subproblems as follows:

(a) $\begin{cases} v_{tt} = \Delta v + te^{5x} \sin(3y) \cos(4z), & -\infty < x, y, z < \infty, \quad t > 0, \\ v|_{t=0} = 0, \quad v_t|_{t=0} = 0, & -\infty < x, y, z < \infty, \end{cases}$

and

(b) $\begin{cases} w_{tt} = a^2\Delta w, & -\infty < x, y, z < \infty, \qquad t > 0, \\ w|_{t=0} = e^{6x+8y}\cos(10z), \quad w_t|_{t=0} = e^{3y+4z}\sin(5x), \quad -\infty < x, y, z < \infty. \end{cases}$

It is readily verified that the solution of problem (a) is the function

$$v(x, y, z, t) = \frac{1}{6} t^3 e^{5x} \sin(3y) \cos(4z).$$

Indeed,

$$\Delta v = \frac{\partial^2 v}{\partial x^2} + \frac{\partial^2 v}{\partial y^2} + \frac{\partial^2 v}{\partial z^2} =$$

$$= \frac{1}{6} t^3 \left[25 e^{5x} \sin(3y)\cos(4z) - 9 e^{5x}\sin(3y)\cos(4z) - 16 e^{5x}\sin(3y)\cos(4z) \right] = 0$$

and

$$v_{tt} = t e^{5x} \sin(3y) \cos(4z).$$

Further, the functions $\varphi(x, y, z) = e^{6x+8y}\cos(10z)$ and $\psi(x, y, z) = e^{3y+4z}\sin(5x)$ are solutions of the Laplace equation (i.e., are harmonic in the whole three-dimensional space), since, for instance,

$$\Delta\varphi = \frac{\partial^2\varphi}{\partial x^2} + \frac{\partial^2\varphi}{\partial y^2} + \frac{\partial^2\varphi}{\partial z^2} =$$

$$= 36 e^{6x+8y}\cos(10z) + 64 e^{6x+8y}\cos(10z) - 100 e^{6x+8y}\cos(10z) = 0.$$

Therefore, the function

$$w(x, y, z, t) = e^{6x+8y}\cos(10z) + t e^{3y+4z}\sin(5x)$$

is a solution of problem (b).

We conclude that the solution of the original problem is

$$u(x, y, z, t) = v(x, y, z, t) + w(x, y, z, t) =$$

$$= \frac{1}{6} t^3 e^{5x}\sin(3y)\cos(4z) + e^{6x+8y}\cos(10z) + e^{3y+4z}\sin(5x).$$

Example 2. Solve the Cauchy problem

$$\begin{cases} u_{tt} = a^2\Delta u, & -\infty < x, y < \infty, \qquad t > 0, \\ u|_{t=0} = \cos(bx + cy), \quad u_t|_{t=0} = \sin(bx + cy), \quad -\infty < x, y < \infty. \end{cases}$$

Solution. Let us decompose the problem into two subproblems as follows:

(1)
$$\begin{cases} u_{tt}^{(1)} = a^2 \Delta u^{(1)}, & -\infty < x, y < \infty, \qquad t > 0, \\ u^{(1)}|_{t=0} = \cos(bx + cy), \quad u_t^{(1)}|_{t=0} = 0, & -\infty < x, y < \infty \end{cases}$$

and

(2)
$$\begin{cases} u_{tt}^{(2)} = a^2 \Delta u^{(2)}, & -\infty < x, y < \infty, \qquad t > 0, \\ u^{(2)}|_{t=0} = 0, \quad u_t^{(2)}|_{t=0} = \sin(bx + cy), & -\infty < x, y < \infty. \end{cases}$$

We will seek the solution of problem (1) in the form

$$u^{(1)}(x, y, t) = \varphi(t) \cos(bx + cy),$$

taking $\varphi(0) = 1$ and $\varphi'(0) = 0$ to ensure that the initial conditions in (1) are satisfied. Substituting this expression of $u^{(1)}$ in the equation $u_{tt}^{(1)} = a^2 \Delta u^{(1)}$ we obtain

$$\varphi'' \cos(bx + cy) = -a^2(b^2 + c^2)\varphi \cos(bx + cy).$$

We thus arrive at the Cauchy problem

$$\begin{cases} \varphi'' + a^2(b^2 + c^2)\varphi = 0, & t > 0, \\ \varphi(0) = 1, \quad \varphi'(0) = 0. \end{cases}$$

Its solution is

$$\varphi(t) = \cos\left(a\sqrt{b^2 + c^2}\, t\right).$$

Hence,

$$u^{(1)}(x, y, t) = \cos(bx + cy) \cos\left(a\sqrt{b^2 + c^2}\, t\right).$$

Similarly, we will seek the solution of problem (2) in the form

$$u^{(2)}(x, y, t) = \psi(t) \sin(bx + cy),$$

taking $\psi(0) = 0$, $\psi'(0) = 1$.

By analogy with problem (1), we are led to the Cauchy problem

$$\begin{cases} \psi'' + a^2(b^2 + c^2)\psi = 0, & t > 0, \\ \psi(0) = 0, \quad \psi'(0) = 1, \end{cases}$$

with the solution

$$\psi(t) = \frac{\sin\left(a\sqrt{b^2 + c^2}\, t\right)}{a\sqrt{b^2 + c^2}}.$$

Hence,

$$u^{(2)}(x, y, t) = \sin(bx + cy) \frac{\sin\left(a\sqrt{b^2 + c^2}\, t\right)}{\left(a\sqrt{b^2 + c^2}\right)}.$$

We conclude that the solution of the original problem is

$$u(x, y, t) = u^{(1)}(x, y, t) + u^{(2)}(x, y, t) =$$

$$= \cos(bx + cy) \cos\left(a\sqrt{b^2 + c^2}\, t\right) + \frac{1}{a\sqrt{b^2 + c^2}} \sin(bx + cy) \sin\left(a\sqrt{b^2 + c^2}\, t\right).$$

2.3. The Fourier integral transform method

The one-dimensional case (one space variable). It is known that if a function $f(x)$ $(-\infty < x < \infty)$ satisfies certain conditions, then the following *Fourier integral formula* holds:

$$f(x) = \frac{1}{2\pi} \int_{-\infty}^{\infty} d\lambda \int_{-\infty}^{\infty} f(\zeta) e^{i\lambda(x-\zeta)}\, d\zeta.$$

By definition, the *Fourier transform* or *image* of the function $f(x)$ is the function

$$\widetilde{f}(\lambda) = \frac{1}{\sqrt{2\pi}} \int_{-\infty}^{\infty} f(x) e^{-i\lambda x}\, dx.$$

Thanks to the Fourier integral formula the function $f(x)$ can be recovered from its Fourier transform via the formula

$$\widetilde{f}(\lambda) = \frac{1}{\sqrt{2\pi}} \int_{-\infty}^{\infty} \widetilde{f}(\lambda) e^{ix\lambda}\, d\lambda.$$

The operation of passing from $f(x)$ to $\widetilde{f}(\lambda)$ is called the *Fourier integral transformation*, whereas that of passing from $\widetilde{f}(\lambda)$ to $f(x)$ is called the *inverse Fourier transformation*.

If the function $f(x)$ is given on the half-line $0 < x < \infty$, then one can consider its *Fourier cosine transform*

$$\widetilde{f}^c(\lambda) = \sqrt{\frac{2}{\pi}} \int_{0}^{\infty} f(x) \cos(\lambda x)\, dx.$$

To pass from $\widetilde{f}^c(\lambda)$ to the original function $f(x)$ one uses the formula

$$f(x) = \sqrt{\frac{2}{\pi}} \int_{0}^{\infty} \widetilde{f}^c(\lambda) \cos(\lambda x)\, d\lambda.$$

Similarly, one defines the *Fourier sine transform*

$$\widetilde{f}^s(\lambda) = \sqrt{\frac{2}{\pi}} \int_{0}^{\infty} f(x) \sin(\lambda x)\, dx,$$

and one has the inversion formula

$$f(x) = \sqrt{\frac{2}{\pi}} \int_{0}^{\infty} \widetilde{f}^s(\lambda) \sin(\lambda x)\, d\lambda.$$

Here are two basic properties of the Fourier transformation.

(1) Linearity: $\widetilde{c_1 f_1 + c_2 f_2} = c_1 \widetilde{f_1} + c_2 \widetilde{f_2}$ (where c_1 and c_2 are arbitrary constants);

(2) Rule of transformation of partial derivatives: if $u = u(x,t)$ and the Fourier transformation is take with respect to the x variable, then

$$\widetilde{u_x} = i\lambda \widetilde{u}, \quad \widetilde{u_{xx}} = (i\lambda)^2 \widetilde{u}, \ldots, \widetilde{u_{\underbrace{x \ldots x}_{n}}} = (i\lambda)^n \widetilde{u}.$$

(this is established by integration by parts). The transforms of the partial derivatives with respect to t are given by

$$\widetilde{u_t} = \frac{\partial \widetilde{u}}{\partial t}, \quad \widetilde{u_{tt}} = \frac{\partial^2 \widetilde{u}}{\partial t^2}, \ldots, \widetilde{u_{\underbrace{t \ldots t}_{n}}} = \frac{\partial^n \widetilde{u}}{\partial t^n}$$

(under the assumption that the indicated partial derivatives with respect to t satisfy certain conditions).

We see that under the action of the Fourier transformation the operation of differentiation with respect to x becomes the operation of multiplication (by $i\lambda$). This important fact is used in solving boundary value problems for partial differential equations.

The Fourier transformation enjoys many other interesting properties which will not be discussed here.

Examples of Fourier transforms are given in Table 2.1.

Table 2.1 Fourier transforms of several functions		
no.	Function $f(x)$	Function $\widetilde{f}(\lambda)$
1	$\begin{cases} 1, & \text{if } \|x\| \leq a, \\ 0, & \text{if } \|x\| > a. \end{cases}$	$\sqrt{\dfrac{2}{\pi}} \dfrac{\sin(\lambda a)}{\lambda}$
2	e^{-ax^2}	$\dfrac{1}{\sqrt{2a}} e^{-\frac{\lambda^2}{4a}}$
3	$e^{-a\|x\|}$	$\sqrt{\dfrac{2}{\pi}} \dfrac{a}{a^2 + \lambda^2}$
4	$\dfrac{1}{x^2 + a^2}$	$\dfrac{1}{2}\sqrt{\dfrac{\pi}{2}} e^{-a\|\lambda\|}$
5	$f(ax)$	$\dfrac{1}{a}\widetilde{f}\left(\dfrac{\lambda}{a}\right)$
6	$f(x - a)$	$e^{-ia\lambda}\widetilde{f}(\lambda)$

Let us return to the boundary value problems for the one-dimensional wave equation. To solve such a problem for $u(x,t)$, one uses the Fourier transformation with respect to the x variable to pass to a problem for the transform $\widetilde{u}(\lambda, t)$. After we solve the latter, one finds the function $u(x,t)$ by applying the inverse Fourier transformation (Figure 2.9).

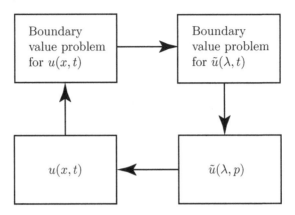

FIGURE 2.9. Block diagram of the Fourier transform method

Example 1 [4, Ch. II, no. 178]. Solve the boundary value problem

$$\begin{cases} u_{tt} = a^2 u_{xx}, & 0 < x < \infty, \qquad 0 < t < \infty, \\ u_x(x,0) = 0, & 0 \le t < \infty, \\ u(x,0) = f(x), & u_t(x,0) = g(x), \qquad 0 \le x < \infty. \end{cases}$$

Solution. Let us apply the Fourier cosine transform with respect to x:

$$\widetilde{u}^c(\lambda, t) = \sqrt{\frac{2}{\pi}} \int_0^\infty u(\zeta, t) \cos(\lambda \zeta) \, d\zeta.$$

We can immediately verify that the boundary condition $u_x(0,t) = 0$ is satisfied. Indeed,

$$u(x,t) = \sqrt{\frac{2}{\pi}} \int_0^\infty \widetilde{u}^c(\lambda, t) \cos(\lambda x) \, d\lambda,$$

whence, upon differentiating with respect to x,

$$u_x(x,t) = -\sqrt{\frac{2}{\pi}} \lambda \int_0^\infty \widetilde{u}^c(\lambda, t) \sin(\lambda x) \, d\lambda,$$

which implies $u_x(0,t) = 0$. Next, using the relations

$$\widetilde{u}_{tt}^c(\lambda, t) = \sqrt{\frac{2}{\pi}} \int_0^\infty u_{tt}(\zeta, t) \cos(\lambda \zeta) \, d\zeta,$$

$$-\lambda^2 \widetilde{u}^c(\lambda, t) = \sqrt{\frac{2}{\pi}} \int_0^\infty u_{\zeta\zeta}(\zeta, t) \cos(\lambda\zeta) \, d\zeta,$$

$$\widetilde{f}^c(\lambda) = \sqrt{\frac{2}{\pi}} \int_0^\infty f(\zeta) \cos(\lambda\zeta) \, d\zeta,$$

and

$$\widetilde{g}^c(\lambda) = \sqrt{\frac{2}{\pi}} \int_0^\infty g(\zeta) \cos(\lambda\zeta) \, d\zeta,$$

we are led to the Cauchy problem for a second-order ordinary differential equation:

$$\begin{cases} \dfrac{d^2 \widetilde{u}^c(\lambda, t)}{dt^2} + a^2 \lambda^2 \widetilde{u}^c(\lambda, t) = 0, & t > 0, \\ \widetilde{u}^c(\lambda, 0) = \widetilde{f}^c(\lambda), & \widetilde{u}_t^c(\lambda, 0) = \widetilde{g}^c(\lambda), \end{cases}$$

with the unknown function $\widetilde{u}^c(\lambda, t)$. The solution of this equation has the form

$$\widetilde{u}^c(\lambda, t) = \widetilde{f}^c(\lambda) \cos(a\lambda t) + \widetilde{g}^c(\lambda) \frac{\sin(a\lambda t)}{a\lambda}.$$

Now the sought function $u(x, t)$ can be found by applying the inverse Fourier cosine transform:

$$u(x, t) = \sqrt{\frac{2}{\pi}} \int_0^\infty \widetilde{u}^c(\lambda, t)) \cos(\lambda x) \, d\lambda =$$

$$= \sqrt{\frac{2}{\pi}} \int_0^\infty \left[\widetilde{f}^c(\lambda) \cos(a\lambda t) + \widetilde{g}^c(\lambda) \frac{\sin(a\lambda t)}{a\lambda} \right] \cos(\lambda x) d\lambda =$$

$$= \sqrt{\frac{2}{\pi}} \int_0^\infty \widetilde{f}^c(\lambda) \cos(a\lambda t) \cos(\lambda x) \, d\lambda + \sqrt{\frac{2}{\pi}} \int_0^\infty \widetilde{g}^c(\lambda) \frac{\sin(a\lambda t)}{a\lambda} \cos(\lambda x) d\lambda.$$

Let us consider separately two cases:

(1) $x - at > 0$. Then

$$u(x, t) = \frac{1}{2} \sqrt{\frac{2}{\pi}} \int_0^\infty \widetilde{f}^c(\lambda) [\cos\lambda(x + at) + \cos\lambda(x - at)] d\lambda +$$

$$+ \frac{1}{2a} \sqrt{\frac{2}{\pi}} \int_0^\infty \widetilde{g}^c(\lambda) \frac{\sin\lambda(x + at) - \sin\lambda(x - at)}{\lambda} \, d\lambda =$$

$$= \frac{f(x + at) + f(x - at)}{2} + \frac{1}{2a} \sqrt{\frac{2}{\pi}} \int_0^\infty \widetilde{g}^c(\lambda) \left[\int_{x-at}^{x+at} \cos(\lambda s) ds \right] d\lambda =$$

$$= \frac{f(x + at) + f(x - at)}{2} + \frac{1}{2a} \int_{x-at}^{x+at} g(s) ds.$$

(2) $x - at < 0$. Then

$$u(x,t) = \frac{1}{2}\sqrt{\frac{2}{\pi}}\int_0^\infty \widetilde{f}^c(\lambda)[\cos\lambda(x+at) + \cos\lambda(at-x)]d\lambda +$$

$$+ \frac{1}{2a}\sqrt{\frac{2}{\pi}}\int_0^\infty \widetilde{g}^c(\lambda)\frac{\sin\lambda(x+at) - \sin\lambda(at-x)}{\lambda}d\lambda =$$

$$= \frac{f(x+at) + f(at-x)}{2} + \frac{1}{2a}\sqrt{\frac{2}{\pi}}\int_0^\infty \widetilde{g}^c(\lambda)\left[\int_{x-at}^{x+at}\cos(\lambda s)ds\right]d\lambda.$$

Since the function $g(x)$ is defined only for positive values of x, the last integral must be transformed as follows:

$$\int_{x-at}^{x+at}\cos(\lambda s)ds = \int_{x-at}^0 \cos(\lambda s)ds + \int_0^{x+at}\cos(\lambda s)ds =$$

$$-\int_{at-x}^0 \cos(\lambda s)ds + \int_0^{x+at}\cos(\lambda s)ds =$$

$$= \int_0^{at-x}\cos(\lambda s)ds + \int_0^{x+at}\cos(\lambda s)ds.$$

Therefore,

$$\sqrt{\frac{2}{\pi}}\int_0^\infty \widetilde{g}^c(\lambda)\left[\int_{x-at}^{x+at}\cos(\lambda s)ds\right]d\lambda =$$

$$= \left[\int_0^{x+at}ds\left(\sqrt{\frac{2}{\pi}}\int_0^\infty \widetilde{g}^c(\lambda)\cos(\lambda s)\,d\lambda\right) + \right.$$

$$+ \left.\int_0^{at-x}ds\left(\sqrt{\frac{2}{\pi}}\int_0^\infty \widetilde{g}^c(\lambda)\cos(\lambda s)\,d\lambda\right)\right] =$$

$$= \int_0^{x+at}g(s)ds + \int_0^{at-x}g(s)ds.$$

We conclude that the sought solution is

$$u(x,t) = \frac{f(x+at) + f(x-at)}{2} + \frac{1}{2a}\left[\int_0^{x+at}g(s)ds - \text{sign}(x-at)\int_0^{|x-at|}g(s)ds\right],$$

where

$$\text{sign} = \begin{cases} 1, & \text{if } x > 0, \\ 0, & \text{if } x = 0, \\ -1, & \text{if } x < 0. \end{cases}$$

Remark 1. For the functions $f(x)$ and $g(x)$ one could take, for example, the functions e^{-ax^2}, e^{-ax} $(x > 0)$, $1/(a^2 + x^2)$, etc.

Remark 2. To solve the problem on the half-line with the boundary condition $u(0,t) = 0$ we must use the Fourier sine transform. Indeed, if

$$\widetilde{u}^{\mathrm{s}}(\lambda, t) = \sqrt{\frac{2}{\pi}} \int_0^\infty u(\zeta, t) \sin(\lambda\zeta) \, d\zeta,$$

then

$$u(x, t) = \sqrt{\frac{2}{\pi}} \int_0^\infty \widetilde{u}^{\mathrm{s}}(\lambda, t) \sin(\lambda x) \, d\lambda,$$

whence $u(0,t) = 0$.

Now let us consider a more complicated problem in which the Fourier transformation applies.

Example 2 [4, Ch. II, no. 176]. Solve the Cauchy problem

$$\begin{cases} u_{tt} = a^2 u_{xx} + c^2 u + f(x,t), & -\infty < x < \infty, \quad t > 0, \\ u(x,0) = 0, \quad u_t(x,0) = 0, & -\infty < x < \infty. \end{cases}$$

Solution. Let us apply the Fourier transformation with respect to the variable x to both sides of the above partial differential equation. This yields a nonhomogeneous second-order ordinary differential equation with nonzero initial data:

$$\begin{cases} \widetilde{u}_{tt}(\zeta, t) + (a^2\zeta^2 - c^2)\widetilde{u}(\zeta, t) = \widetilde{f}(\zeta, t), & t > 0, \\ \widetilde{u}(\zeta, 0) = \widetilde{u}_t(\zeta, 0) = 0. \end{cases}$$

To solve this Cauchy problem we will use Duhamel's method. Specifically, let us solve the auxiliary Cauchy problem

$$\begin{cases} v_{tt} + (a^2\zeta^2 - c^2)v = 0, & v = v(\zeta, t, \tau), \quad t > \tau, \\ v|_{t=\tau} = 0, \quad v_t|_{t=\tau} = \widetilde{f}(\zeta, \tau). \end{cases}$$

Then the solution $\widetilde{u}(\zeta, t)$ of the original (Fourier-transformed) problem is given by the formula

$$\widetilde{u}(\zeta, t) = \int_0^t v(\zeta, t, \tau) \, d\tau,$$

where

$$v(\zeta, t, \tau) = C_1 \cos\left(\sqrt{a^2\zeta^2 - c^2}\, t\right) + C_2 \sin\left(\sqrt{a^2\zeta^2 - c^2}\, t\right).$$

This yields

$$\begin{cases} 0 = C_1 \cos\left(\sqrt{a^2\zeta^2 - c^2}\, \tau\right) + C_2 \sin\left(\sqrt{a^2\zeta^2 - c^2}\, \tau\right), \\ \widetilde{f}(\zeta, \tau) = -C_1 \sqrt{a^2\zeta^2 - c^2} \sin\left(\sqrt{a^2\zeta^2 - c^2}\, \tau\right) + \\ \qquad + C_2 \sqrt{a^2\zeta^2 - c^2} \cos\left(\sqrt{a^2\zeta^2 - c^2}\, \tau\right). \end{cases}$$

Denote $\sqrt{a^2\zeta^2 - c^2} = b$. Then

$$\begin{cases} C_1 \cos(b\tau) + C_2 \sin(b\tau) = 0, \\ C_1(-\sin(b\tau)) + C_2 \cos(b\tau) = \dfrac{\widetilde{f}(\zeta,\tau)}{b}, \end{cases}$$

whence

$$C_1 = \begin{vmatrix} 0 & \sin(b\tau) \\ \frac{\widetilde{f}(\zeta,\tau)}{b} & \cos(b\tau) \end{vmatrix} = -\frac{\widetilde{f}(\zeta,\tau)}{b} \sin(b\tau)$$

and

$$C_2 = \begin{vmatrix} \cos(b\tau) & 0 \\ -\sin(b\tau) & \frac{\widetilde{f}(\zeta,\tau)}{b} \end{vmatrix} = \frac{\widetilde{f}(\zeta,\tau)}{b} \cos(b\tau).$$

Therefore,

$$v(\zeta,t,\tau) = -\frac{\widetilde{f}(\zeta,\tau)}{b} \sin(b\tau)\cos(bt) + \frac{\widetilde{f}(\zeta,\tau)}{b} \cos(b\tau)\sin(bt) =$$

$$= \frac{\widetilde{f}(\zeta,\tau)}{b} \sin b(t-\tau) = \frac{\widetilde{f}(\zeta,\tau)}{\sqrt{a^2\zeta^2 - c^2}} \sin\left(\sqrt{a^2\zeta^2 - c^2}(t-\tau)\right),$$

i.e.,

$$\widetilde{u}(\zeta,t) = \int_0^t \widetilde{f}(\zeta,\tau)\frac{\sin\left(\sqrt{a^2\zeta^2 - c^2}(t-\tau)\right)}{\sqrt{a^2\zeta^2 - c^2}} \, d\tau.$$

Now applying the formula for the inverse Fourier transform we get

$$u(x,t) = \frac{1}{\sqrt{2\pi}} \int_{-\infty}^{\infty} \widetilde{u}(\zeta,t)e^{i\zeta x} \, d\zeta =$$

$$= \frac{1}{\sqrt{2\pi}} \int_0^t d\tau \int_{\infty}^{\infty} \widetilde{f}(\zeta,\tau)\frac{\sin\left(\sqrt{a^2\zeta^2 - c^2}(t-\tau)\right)}{\sqrt{a^2\zeta^2 - c^2}} e^{i\zeta x} \, d\zeta =$$

$$= \frac{1}{\sqrt{2\pi}} \int_0^t d\tau \int_{-\infty}^{\infty} d\zeta \int_{-\infty}^{\infty} f(\lambda,\tau)\frac{\sin\left(\sqrt{a^2\zeta^2 - c^2}(t-\tau)\right)}{\sqrt{a^2\zeta^2 - c^2}} e^{i\zeta(x-\lambda)} d\lambda.$$

In the theory of Bessel function one has the formula

$$\frac{\sin r}{r} = \frac{1}{2} \int_0^\pi J_0(r\sin\varphi\sin\theta)e^{ir\cos\varphi\cos\theta} \sin\theta \, d\theta.$$

Now let us change of variable in this identity according to the rule

$$r\cos\varphi = -a\zeta(t-\tau), \qquad r\sin\varphi = ic(t-\tau).$$

Then $r^2 = (t-\tau)^2(a^2\zeta^2 - c^2)$.

It follows that

$$\frac{\sin\left(\sqrt{a^2\zeta^2 - c^2}(t - \tau)\right)}{\sqrt{a^2\zeta^2 - c^2}(t - \tau)} = \frac{1}{2}\int_0^\pi J_0(ic(t - \tau)\sin\theta)e^{-ia\zeta(t-\tau)\cos\theta}\sin\theta.$$

Next, let us make the change of variable $\cos\theta = \beta/(a(t-\tau))$ in this integral. Then $-\sin\theta d\theta = d\beta/[a(t - \tau)]$ and

$$= \frac{1}{2}\int_0^\pi J_0(ic(t - \tau)\sin\theta)e^{-ia\zeta(t-\tau)\cos\theta}\sin\theta =$$

$$= \frac{1}{2a}\int_{-a(t-\tau)}^{a(t-\tau)} J_0\left(ic(t - \tau)\sqrt{1 - \frac{\beta^2}{a^2(t - \tau)^2}}\right)\frac{e^{-i\zeta\beta}}{t - \tau}d\beta =$$

$$= \frac{1}{2a}\int_{-a(t-\tau)}^{a(t-\tau)} I_0\left(c\sqrt{(t - \tau)^2 - \frac{\beta^2}{a^2}}\right)\frac{e^{-i\zeta\beta}}{t - \tau}d\beta.$$

Consequently

$$u(x,t) = \frac{1}{4\pi a}\int_0^t d\tau \int_{-\infty}^\infty d\zeta \times$$

$$\times \int_{-\infty}^\infty \left[f(\lambda, \tau)\int_{-a(t-\tau)}^{a(t-\tau)} I_0\left(c\sqrt{(t - \tau)^2 - \frac{\beta^2}{a^2}}\right)e^{i\zeta(x - \lambda - \beta)}d\beta\right]d\lambda.$$

Let us set

$$g(\beta) = \begin{cases} I_0\left(c\sqrt{(t - \tau)^2 - \frac{\beta^2}{a^2}}\right), & \text{if } |\beta| \leq a|t - \tau|, \\ 0, & \text{if } |\beta| \geq a|t - \tau|. \end{cases}$$

Then

$$\int_{-a(t-\tau)}^{a(t-\tau)} I_0\left(c\sqrt{(t - \tau)^2 - \frac{\beta^2}{a^2}}\right)e^{i\zeta(x - \beta)}d\beta = \int_{-a(t-\tau)}^{a(t-\tau)} g(\beta)e^{i\zeta(x - \beta)}d\beta.$$

By the Fourier integral formula,

$$\frac{1}{2\pi}\int_{-\infty}^\infty d\zeta \int_{-\infty}^\infty g(\beta)e^{i\zeta(x - \lambda - \beta)}d\beta = g(x - \lambda).$$

This yields (by the definition of the function $g(\beta)$)

$$g(x - \lambda) = \begin{cases} I_0\left(c\sqrt{(t - \tau)^2 - \frac{(x-\lambda)^2}{a^2}}\right), & \text{if } x - a(t - \tau) < \lambda < x + a(t - \tau), \\ 0, & \text{if } -\infty < \lambda < x - a(t - \tau), \\ & \text{or } x + a(t - \tau) < \lambda < \infty. \end{cases}$$

We finally conclude that

$$u(x,t) = \frac{1}{2a}\int_0^\tau d\tau \int_{x-a(t-\tau)}^{x+a(t-\tau)} f(\lambda, \tau)I_0\left(c\sqrt{(t - \tau)^2 - \frac{(x - \lambda)^2}{a^2}}\right)d\lambda.$$

The two-dimensional case (two space variables).

Example 3 [4, Ch. VI, no. 106]. Solve the boundary value problem

$$\begin{cases} u_{tt} = a^2 \left(u_{xx} + u_{yy}\right) + f(x,y,t), & -\infty < x, y < \infty, \quad t > 0, \\ u|_{t=0} = 0, \qquad u_t|_{t=0} = 0, & -\infty < x, y < \infty. \end{cases}$$

Solution. Let us the Fourier transformation with respect to the variables x and y to our equation and the initial condition, using the relations

$$\tilde{u}(\lambda, \mu, t) = \frac{1}{2\pi} \int\!\!\!\int\limits_{-\infty}^{\infty} e^{-i(\lambda x + \mu y)} u(x, y, t) \, dx \, dy$$

$$u(x, y, t) = \frac{1}{2\pi} \int\!\!\!\int\limits_{-\infty}^{\infty} e^{i(\lambda x + \mu y)} \tilde{u}(\lambda, \mu, t) \, d\lambda \, d\mu$$

$$\tilde{f}(\lambda, \mu, t) = \frac{1}{2\pi} \int\!\!\!\int\limits_{-\infty}^{\infty} e^{-i(\lambda x + \mu y)} f(x, y, t) \, dx \, dy$$

and

$$\int\!\!\!\int\limits_{-\infty}^{\infty} e^{-i(\lambda x + \mu y)} \left(u_{xx} + u_{yy}\right) \, dx \, dy = -(\lambda^2 + \mu^2)\tilde{u}(\lambda, \mu, t)$$

(the last of which is established by means of integration by parts and using the fact that the function $u(x, y, t)$ and its partial derivatives u_x and u_y vanish when x and y become infinite). This yields the Cauchy problem

$$\begin{cases} \tilde{u}_{tt}(\lambda, \mu, t) + a^2 \left(\lambda^2 + \mu^2\right) \tilde{u}(\lambda, \mu, t) = \tilde{f}(\lambda, \mu, t), & t > 0, \\ \tilde{u}(\lambda, \mu, 0) = 0, \qquad \tilde{u}_t(\lambda, \mu, 0) = 0. \end{cases}$$

Solving this new problem by Duhamel's formula (method of variation of constants), we obtain

$$\tilde{u}(\lambda, \mu, t) = \int_0^t \tilde{f}(\lambda, \mu, t) \frac{\sin\left(a(t-\tau)\sqrt{\lambda^2 + \mu^2}\right)}{a\sqrt{\lambda^2 + \mu^2}} \, d\tau.$$

Next, applying the inverse Fourier transformation we have

$$u(x, y, t) = \frac{1}{2\pi} \int_0^t d\tau \times \int\!\!\!\int\limits_{-\infty}^{\infty} \tilde{f}(\lambda, \mu, t) \frac{\sin\left(a(t-\tau)\sqrt{\lambda^2 + \mu^2}\right)}{a\sqrt{\lambda^2 + \mu^2}} e^{i(\lambda x + \mu y)} \, d\lambda \, d\mu.$$

Inserting here the value of $\widetilde{f}(\lambda, \mu, t)$ and denoting $\nu = \sqrt{\lambda^2 + \mu^2}$, we arrive at the relation

$$u(x, y, t) = \frac{1}{(2\pi)^2} \int_0^t d\tau \times$$

$$\times \iiiint_{-\infty}^{\infty} f(\zeta, \eta, \tau) \frac{\sin(a\nu(t - \tau))}{a\nu} e^{i[\lambda(x - \zeta) + \mu(y - \eta)]} \, d\zeta \, d\eta \, d\lambda \, d\mu.$$

To calculate the quadruple indefinite integral we introduce polar coordinates via the relations

$$\begin{cases} \zeta - x = \rho \cos \varphi, \\ \eta - y = \rho \sin \varphi, \end{cases} \qquad \begin{cases} \lambda = \nu \cos \theta, \\ \mu = \nu \sin \theta. \end{cases}$$

This immediately gives

$$\lambda(\zeta - x) + \mu(\eta - y) = \rho\nu \cos(\theta - \varphi) = \rho\nu \cos \psi$$

where $\psi = \theta - \varphi$. It follows that

$$\iiiint_{-\infty}^{\infty} f(\zeta, \eta, \tau) \frac{\sin(a\nu(t - \tau))}{a\nu} e^{i[\lambda(x - \zeta) + \mu(y - \eta)]} \, d\zeta \, d\eta \, d\lambda \, d\mu =$$

$$= \int_0^\infty d\rho \int_0^\infty d\nu \int_0^{2\pi} d\varphi \int_0^{2\pi} f(\zeta, \eta, \tau) \frac{\sin(a\nu(t - \tau))}{a\nu} e^{-i\rho\nu \cos \psi} \rho\nu \, d\psi.$$

Now observe that the following relation (integral representation of the Bessel functions) holds:

$$J_n(x) = \frac{(-i)^n}{2\pi} \int_{-\pi}^{\pi} e^{ix \cos \psi + in\psi} \, d\psi.$$

In particular,

$$J_0(x) = \frac{1}{2\pi} \int_0^{2\pi} e^{ix \cos \psi} \, d\psi,$$

or (considering x real)

$$J_0(x) = \frac{1}{2\pi} \int_0^{2\pi} e^{-ix \cos \psi} \, d\psi.$$

It follows that

$$\int_0^\infty d\rho \int_0^\infty d\nu \int_0^{2\pi} d\varphi \int_0^{2\pi} f(\zeta, \eta, \tau) \frac{\sin(a\nu(t - \tau))}{a\nu} e^{-i\rho\nu \cos \psi} \rho\nu \, d\psi =$$

$$= \frac{2\pi}{a} \int_0^\infty d\rho \int_0^\infty d\nu \int_0^{2\pi} f(\zeta, \eta, \tau) \sin(a\nu(t - \tau)) J_0(\rho\nu) \, d\varphi.$$

We shall also use the fact that

$$
\int_0^\infty J_0(\rho\nu)\sin(a\nu(t-\tau))\,d\nu =
\begin{cases}
0, & \text{if } a(t-\tau) < \rho, \\[2ex]
\dfrac{1}{\sqrt{a^2(t-\tau)^2-\rho^2}}, & \text{if } a(t-\tau) > \rho.
\end{cases}
$$

Therefore,

$$
\int_0^\infty d\rho \int_0^\infty d\nu \int_0^{2\pi} f(\zeta,\eta,\tau)\sin(a\nu(t-\tau))J_0(\rho\nu)\,d\varphi =
$$

$$
= \int_0^{a(t-\tau)} \int_0^{2\pi} \frac{f(\zeta,\eta,\tau)\rho\,d\rho\,d\varphi}{\sqrt{a^2(t-\tau)^2-\rho^2}}.
$$

Finally,

$$
u(x,y,t) = \frac{1}{2\pi a}\int_0^t d\tau \int_0^{a(t-\tau)} \int_0^{2\pi} \frac{f(\zeta,\eta,\tau)\rho\,d\rho\,d\varphi}{\sqrt{a^2(t-\tau)^2-\rho^2}},
$$

or

$$
u(x,y,t) = \frac{1}{2\pi a}\int_0^t d\tau \iint_{\rho\le a(t-\tau)} \frac{f(\zeta,\eta,\tau)\,d\zeta\,d\eta}{\sqrt{a^2(t-\tau)^2-\rho^2}}.
$$

The three-dimensional case (three space variables).

Example 4 [4, Ch. IV, no. 107]. Solve the boundary value problem

$$
\begin{cases}
u_{tt} = a^2\left(u_{xx}+u_{yy}+u_{zz}\right) + f(x,y,z,t), & -\infty < x,y,z < \infty, \quad t > 0, \\
u|_{t=0} = 0, \quad u_t|_{t=0} = 0, & -\infty < x,y,z < \infty.
\end{cases}
$$

Solution. Applying the Fourier transformation with respect to the variables x, y, z to the equation and initial conditions, we obtain the Cauchy problem

$$
\begin{cases}
\widetilde{u}_{tt}(\lambda,\mu,\nu,t) + a^2\left(\lambda^2+\mu^2+\nu\right)\widetilde{u}(\lambda,\mu,\nu,t) = \widetilde{f}(\lambda,\mu,\nu,t), & t > 0, \\
\widetilde{u}(\lambda,\mu,\nu,t)|_{t=0} = 0, \quad \widetilde{u}_t(\lambda,\mu,\nu,t)|_{t=0} = 0,
\end{cases}
$$

where

$$
\widetilde{u}(\lambda,\mu,\nu,t) = \frac{1}{(2\pi)^{3/2}}\int\!\!\!\int\!\!\!\int_{-\infty}^{\infty} e^{-i(\lambda x+\mu y+\nu z)}u(x,y,z,t)\,dx\,dy\,dz,
$$

and

$$
\widetilde{f}(\lambda,\mu,\nu,t) = \frac{1}{(2\pi)^{3/2}}\int\!\!\!\int\!\!\!\int_{-\infty}^{\infty} e^{-i(\lambda x+\mu y+\nu z)}f(x,y,z,t)\,dx\,dy\,dz.
$$

The solution of this problem (by analogy with the preceding one) is given by the integral

$$\tilde{u}(\lambda, \mu, \nu, t) = \int_0^t \tilde{f}(\lambda, \mu, \nu, t) \frac{\sin(a\rho(t - \tau))}{a\rho} d\tau,$$

where $\rho^2 = \lambda^2 + \mu^2 + \nu^2$.

By the formula for the inverse Fourier transformation,

$$u(x, y, z, t) = \frac{1}{(2\pi)^3} \int_0^t d\tau \int \cdots \int f(\xi, \eta, \zeta, \tau) \frac{\sin(a\rho(t - \tau))}{a\rho} \times$$

$$\times e^{i[\lambda(x-\xi)+\mu(y-\eta)+\nu(z-\zeta)]} d\xi\, d\eta\, d\zeta\, d\lambda\, d\mu\, d\nu,$$

where the sextuple integral is taken over the domain $\{-\infty < \xi, \eta, \zeta, \lambda\, \mu, \nu < \infty\}$. To calculate this integral we will pass to polar coordinates via the formulas

$$\begin{cases} \xi - x = r\sin\theta\cos\varphi, & \eta - y = r\sin\theta\sin\varphi, & \zeta - z = r\cos\theta, \\ \lambda = \rho\sin\sigma\cos\psi, & \mu = \rho\sin\sigma\sin\psi, & \nu = \rho\cos\sigma,, \end{cases}$$

where σ is the angle between the vectors $\vec{r} = (\xi - x)\vec{i} + (\eta - y)\vec{j} + (\zeta - z)\vec{k}$ and $\vec{\rho} = \lambda\vec{i} + \mu\vec{j} + \nu\vec{k}$. Then we have

$$u(x, y, z, t) = \frac{1}{(2\pi)^3} \int_0^t d\tau \int_0^\infty \int_0^\infty \int_0^\pi \int_0^\pi \int_0^{2\pi} \int_0^{2\pi} \times$$

$$\times f(x + r\sin\theta\cos\varphi, y + r\sin\theta\sin\varphi, z + r\cos\theta, \tau) \times$$

$$\times \frac{\sin(a\rho(t - \tau))}{a\rho} \rho^2 r^2 \sin\theta \sin\sigma\, e^{-i\rho r\cos\sigma} d\rho\, dr\, d\theta\, d\sigma\, d\varphi\, d\psi =$$

$$= \frac{1}{(2\pi)^3 a} \int_0^t d\tau \int_0^\infty \int_0^\infty \int_0^\pi \int_0^{2\pi} f(x + r\sin\theta\cos\varphi, y + r\sin\theta\sin\varphi, z + r\cos\theta, \tau) \times$$

$$\times \sin(a\rho(t - \tau))\rho r^2 \sin\theta\, d\rho\, dr\, d\theta\, d\varphi \int_0^\pi e^{-i\rho r\cos\sigma} \sin\sigma\, d\sigma \int_0^{2\pi} d\psi.$$

Observing that

$$\int_0^\pi e^{-i\rho r\cos\sigma} \sin\sigma\, d\sigma = \frac{2\sin(\rho r)}{\rho r}, \qquad \int_0^{2\pi} d\psi = 2\pi,$$

we get

$$u(x, y, z, t) = \frac{1}{2\pi^2 a} \int_0^t d\tau \times$$

$$\times \int_0^\infty \int_0^\infty \int_0^\pi \int_0^{2\pi} f(x + r\sin\theta\cos\varphi, y + r\sin\theta\sin\varphi, z + r\cos\theta, \tau) \times$$

$$\times \sin(a\rho(t - \tau))\sin(\rho r)\sin\theta\, r d\rho\, dr\, d\theta\, d\varphi =$$

$$= \frac{1}{4\pi^2 a} \int_0^t d\tau \int_0^\infty \int_0^\infty \int_0^\pi \int_0^{2\pi} f(x + r \sin\theta \cos\varphi, y + r \sin\theta \sin\varphi, z + r \cos\theta, \tau) \times$$

$$\times r[\cos\rho(r - a(t - \tau)) - \cos\rho(r + a(t - \tau))] \sin\theta \, d\rho \, dr \, d\theta \, d\varphi.$$

Now note that the following equality holds (Fourier integral):

$$\frac{1}{\pi} \int_0^\infty d\rho \int_0^\infty g(r) \cos(\rho(r - s)) \, dr = g(s),$$

whenever $g(r) = 0$ for $r \leq 0$.

Hence, choosing

$$g(r) = \begin{cases} rf(x + r \sin\theta \cos\varphi, y + r \sin\theta \sin\varphi, z + r \cos\theta, \tau) & \text{if } r \geq 0, \\ 0 & \text{if } r \leq 0, \end{cases}$$

we get

$$\frac{1}{\pi} \int_0^\infty d\rho \int_0^\infty rf(x + r \sin\theta \cos\varphi, y + r \sin\theta \sin\varphi, z + r \cos\theta, \tau) \times$$

$$\times \cos[\rho(r - a(t - \tau))] \, dr =$$

$$= a(t - \tau)f(x + a(t - \tau) \sin\theta \cos\varphi, y + a(t - \tau) \sin\theta \sin\varphi, z + a(t - \tau) \cos\theta, \tau).$$

Similarly, the function

$$\frac{1}{\pi} \int_0^\infty d\rho \int_0^\infty rf(x + r \sin\theta \cos\varphi, y + r \sin\theta \sin\varphi, z + r \cos\theta, \tau) \times$$

$$\times \cos[\rho(r + a(t - \tau))] \, dr$$

is equal to

$$[-a(t - \tau)]f(x - a(t - \tau) \sin\theta \cos\varphi, y - a(t - \tau) \sin\theta \sin\varphi, z - a(t - \tau) \cos\theta, \tau)$$

if $t \leq \tau$, and is equal to 0 if $t \geq \tau$. We conclude that

$$u(x, y, z, t) = \frac{a}{4\pi} \int_0^t (t - \tau) \, d\tau \int_0^\pi \int_0^{2\pi} \times$$

$$\times f(x + a(t - \tau) \sin\theta \cos\varphi, y + a(t - \tau) \sin\theta \sin\varphi, z + a(t - \tau) \cos\theta, \tau) \sin\theta \, d\theta \, d\varphi.$$

Now subjecting the last integral to the change of variables $r = a(t - \tau)$ (whence $\tau = t - r/a$) and passing from the surface integral to a volume integral, we finally obtain

$$u(x, y, z, t) = \frac{1}{4\pi a^2} \iiint_{r \leq at} \frac{f(\xi, \eta, \zeta, t - r/a)}{r} \, d\xi \, d\eta \, d\zeta,$$

where $r = \sqrt{(x - \xi)^2 + (y - \eta)^2 + (z - \zeta)^2}$.

To conclude this section we will consider yet another example of application of the Fourier transformation.

Example 5 [4, Ch. IV, no. 108]. Solve the following boundary value problem (on the oscillations of an infinite plate):

$$\begin{cases} u_{tt} + b^2 \Delta^2 u = 0, & -\infty < x, y < \infty, \quad t > 0, \\ u|_{t=0} = f(x,y), & u_t|_{t=0} = 0, \quad -\infty < x, y < \infty. \end{cases}$$

Solution. Let us apply the Fourier transformation with respect to the space variables to the given equation and initial conditions. First we find

$$\iint\limits_{-\infty}^{\infty} \Delta^2 u \, e^{-i(\lambda x + \mu y)} \, dx \, dy = \iint\limits_{-\infty}^{\infty} \left(\frac{\partial^4 u}{\partial x^4} + 2\frac{\partial^4 u}{\partial x^2 \partial y^2} + \frac{\partial^4 u}{\partial y^4} \right) e^{-i(\lambda x + \mu y)} \, dx \, dy.$$

Integrating by parts (and observing that the terms outside the integral symbol vanish) we get

$$\iint\limits_{-\infty}^{\infty} \Delta^2 u \, e^{-i(\lambda x + \mu y)} \, dx \, dy = (\lambda^2 + \mu^2) \widetilde{u}(\lambda, \mu, t),$$

where, as usual,

$$\widetilde{u}(\lambda, \mu, t) = \frac{1}{2\pi} \iint\limits_{-\infty}^{\infty} u(x,y,t) e^{-i(\lambda x + \mu y)} \, dx \, dy.$$

Also,

$$\widetilde{f}(\lambda, \mu) = \frac{1}{2\pi} \iint\limits_{-\infty}^{\infty} f(x,y) e^{-i(\lambda x + \mu y)} \, dx \, dy.$$

Thus, we arrive at the Cauchy problem

$$\begin{cases} \widetilde{u}_{tt}(\lambda, \mu, t) + b^2(\lambda^2 + \mu^2)\widetilde{u}(\lambda, \mu, t) = 0, & -\infty < x, y < \infty, \quad t > 0, \\ \widetilde{u}|_{t=0} = \widetilde{f}(\lambda, \mu), & \widetilde{u}_t|_{t=0} = 0, \quad -\infty < x, y < \infty. \end{cases}$$

The solution of this problem is

$$\widetilde{u}(\lambda, \mu, t) = \widetilde{f}(\lambda, \mu) \cos(b(\lambda^2 + \mu^2)t).$$

Now using the inverse Fourier transformation we obtain

$$u(x,y,t) = \frac{1}{(2\pi)^2} \iint\limits_{-\infty}^{\infty} \iint\limits_{-\infty}^{\infty} f(\xi, \eta) \cos(bt(\lambda^2 + \mu^2)) e^{i[\lambda(x-\xi) + \mu(y-\eta)]} \, d\xi \, d\eta \, d\lambda \, d\mu =$$

$$= \frac{1}{(2\pi)^2} \iint\limits_{-\infty}^{\infty} \iint\limits_{-\infty}^{\infty} f(\xi, \eta) \cos \lambda(x-\xi) \cos \mu(y-\eta) \cos(bt(\lambda^2 + \mu^2)) \, d\xi \, d\eta \, d\lambda \, d\mu$$

(here we used the formula $e^{i\varphi} = \cos\varphi + i\sin\varphi$ and the fact that the integrals containing $\sin \lambda(x-\xi)$ and $\sin \mu(y-\eta)$ are equal to zero).

Let us carry out the integration with respect to λ and μ using the relations

$$\int_{-\infty}^{\infty} \cos(p\sigma^2)\cos(q\sigma)\,d\sigma = \sqrt{\frac{\pi}{p}}\cos\left(\frac{\pi}{4} - \frac{q^2}{4p}\right),$$

$$\int_{-\infty}^{\infty} \sin(p\sigma^2)\cos(q\sigma)\,d\sigma = \sqrt{\frac{\pi}{p}}\sin\left(\frac{\pi}{4} - \frac{q^2}{4p}\right), \qquad p > 0.$$

This yields

$$\iint\limits_{-\infty}^{\infty} \cos\lambda(x-\xi)\,\cos\mu(y-\eta)\cos(bt(\lambda^2 + \mu^2))]d\lambda\,d\mu =$$

$$= \iint\limits_{-\infty}^{\infty} \cos\lambda(x-\xi)\,\cos\mu(y-\eta)[\cos(bt\lambda^2)\cos(bt\mu^2) - \sin(bt\lambda^2)\sin(bt\mu^2)\,d\lambda\,d\mu =$$

$$= \int_{-\infty}^{\infty} \cos\lambda(x-\xi)\cos(bt\lambda^2)\,d\lambda \cdot \int_{-\infty}^{\infty} \cos\mu(y-\eta)\cos(bt\mu^2)\,d\mu -$$

$$- \int_{-\infty}^{\infty} \sin(bt\lambda^2)\cos\lambda(x-\xi)\,d\lambda \cdot \int_{-\infty}^{\infty} \sin(bt\mu^2)\cos\mu(y-\eta)\,d\mu =$$

$$= \sqrt{\frac{\pi}{bt}}\cos\left(\frac{\pi}{4} - \frac{(x-\xi)^2}{4bt}\right) \cdot \sqrt{\frac{\pi}{bt}}\cos\left(\frac{\pi}{4} - \frac{(y-\eta)^2}{4bt}\right) -$$

$$- \sqrt{\frac{\pi}{bt}}\sin\left(\frac{\pi}{4} - \frac{(x-\xi)^2}{4bt}\right) \cdot \sqrt{\frac{\pi}{bt}}\sin\left(\frac{\pi}{4} - \frac{(y-\eta)^2}{4bt}\right) =$$

$$= \frac{\pi}{bt}\left[\cos\left(\frac{\pi}{4} - \alpha^2\right)\cos\left(\frac{\pi}{4} - \beta^2\right) - \sin\left(\frac{\pi}{4} - \alpha^2\right)\sin\left(\frac{\pi}{4} - \beta^2\right)\right] =$$

$$= \frac{\pi}{bt}\sin(\alpha^2 + \beta^2),$$

where we denote

$$\alpha = \frac{\xi - x}{2\sqrt{bt}}, \qquad \beta = \frac{\eta - y}{2\sqrt{2bt}}.$$

Passing to the variables α and β in the integral

$$\frac{1}{(2\pi)^2}\iint\limits_{-\infty}^{\infty} f(\xi,\eta)\frac{\pi}{bt}\sin(\alpha^2 + \beta^2)\,d\xi\,d\eta$$

we finally obtain

$$u(x,y,t) = \frac{1}{\pi}\iint\limits_{-\infty}^{\infty} f(x + 2\alpha\sqrt{bt}, y + 2\beta\sqrt{bt})\sin(\alpha^2 + \beta^2)\,d\alpha\,d\beta.$$

2.4. The Laplace integral transform method

We have already demonstrated how to the Fourier transformation applies in solving some problems. In this section we will become acquainted with the Laplace transformation.

Let the function $f(t)$ satisfy the following conditions:

(a) $f(t)$ is piecewise continuous on the segment $[0, a]$ for any $a > 0$;

(b) $f(t) = 0$ for $t < 0$;

(c) there exist numbers $M > 0$ and $s_0 > 0$ such that $|f(t)| \le M^{s_0 t}$.

Then the *Laplace transform* of the function $f(t)$ is defined to be

$$L[f(t)] \equiv F(p) = \int_0^\infty e^{-pt} f(t)\, dt,$$

where $p = s + i\sigma$; this will be denoted by $f(t) \doteq F(p)$.

Clearly, the Laplace transformation is defined if $s > s_0$ (this ensures that the integral converges).

Let us indicate two basic properties of the Laplace transformation.

(1) Linearity, i.e., $L[C_1 f_1 + C_2 f_2] = C_1 L[f_1] + C_2 L[f_2]$, where C_1 and C_2 are arbitrary constants.

(2) Partial derivatives transform according to the following rules: if $u = u(x, t)$ and the Laplace transformation is taken with respect to the variable t ($t > 0$), then, denoting

$$L[u(x, t)] \equiv U(x, p) = \int_0^\infty e^{-pt} u(x, t)\, dt,$$

and assuming that certain conditions on the function $u(x, t)$ and its partial derivatives are satisfied, one has

$$L\left[\frac{\partial u}{\partial t}\right] = pU(x, p) - u(x, 0),$$

$$L\left[\frac{\partial^2 u}{\partial t^2}\right] = p^2 U(x, p) - pu(x, 0) + \frac{\partial u}{\partial t}(x, 0),$$

$$L\left[\frac{\partial u}{\partial x}\right] = \frac{\partial U}{\partial x}(x, p), \qquad L\left[\frac{\partial^2 u}{\partial x^2}\right] = \frac{\partial^2 U}{\partial x^2}(x, p).$$

These equalities are established by integration by parts.

The Laplace transformation has of course many other properties, but we will not dwell further upon them.

We see that the Laplace transformation replaces differentiation with respect to the time variable t by multiplication (by p). This important observation is used in solving partial differential equations.

The block diagram of the application of the Laplace transformation to the solution of partial differential equations is completely analogous to that for the Fourier transformation (Figure 2.10).

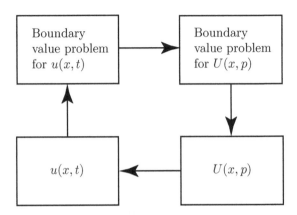

FIGURE 2.10. Block diagram of the Laplace transform method

Table 2.2 shows the Laplace transforms of several functions.

Now let us demonstrate how the Laplace transformation applies in solving hyperbolic equations.

Example 1 [3, no. 383]. Suppose that starting with the moment $t = 0$ an electromotive force $E(t)$ is applied at the endpoint $x = 0$ of a semi-infinite isolated electric line. Find the voltage $u(x,t)$ in the line if it is known that the initial voltage and the initial current are equal to zero, in the following two cases:

(a) lossless line $(R = G = 0)$;

(b) distortionless line $(RC = RL)$.

Solution. The mathematical formulation of problem (a) has the form

$$\text{(a)} \quad \begin{cases} u_{tt} = \dfrac{1}{LC}\, u_{xx}, & 0 < x < \infty, \quad 0 < t < \infty, \\ u(0,t) = E(t), & 0 \le t < \infty, \\ u(x,0) = 0, \quad u_t(x,0) = 0, & 0 \le x < \infty, \end{cases}$$

where L and C stand for the inductance and the capacitance per unit of length of the conductor, respectively.

no.	Function $f(x)$	Function $F(p) = L[f(t)]$
	Table 2.2 Laplace transforms of several functions	
1	Heaviside function $\theta(t)$	$\dfrac{1}{p}$, $\operatorname{Re} p > 0$
2	t^n ($n > 0$ an integer)	$\dfrac{n!}{p^{n+1}}$, $\operatorname{Re} p > 0$
3	e^{at}	$\dfrac{1}{p-a}$, $\operatorname{Re} p > a$
4	$\sin(at)$	$\dfrac{a}{p^2 + a^2}$, $\operatorname{Re} p > 0$
5	$\cos(at)$	$\dfrac{p}{p^2 + a^2}$, $\operatorname{Re} p > 0$
6	$\sinh(at)$	$\dfrac{a}{p^2 - a^2}$, $\operatorname{Re} p > |a|$
7	$\cosh(at)$	$\dfrac{p}{p^2 - a^2}$, $\operatorname{Re} p > |a|$
8	$e^{at}\sin(bt)$	$\dfrac{b}{(p-a)^2 + b^2}$, $\operatorname{Re} p > a$
9	$e^{at}\cos(bt)$	$\dfrac{p-a}{(p-a)^2 + b^2}$, $\operatorname{Re} p > a$
10	$1 - \dfrac{2}{\sqrt{\pi}} \displaystyle\int_0^{a/2\sqrt{z}} e^{-x^2}\,dx$	$\dfrac{1}{p}e^{-a\sqrt{p}}$, $a > 0$
11	$\dfrac{\sin(at)}{t}$	$\dfrac{\pi}{2} - \arctan\dfrac{p}{a}$, $\operatorname{Re} p > 0$
12	$J_0(at)$	$\dfrac{1}{\sqrt{a^2 + p^2}}$

Let us apply the Laplace transformation with respect to t to both sides of our partial differential equation, taking into account that

$$u(x,t) \doteqdot U(x,p), \qquad u_t(x,t) \doteqdot pU(x,p), \qquad u_{tt}(x,t) \doteqdot p^2 U(x,p),$$

and

$$u_{xx}(x,t) \doteqdot U_{xx}(x,p), \qquad E(t) \doteqdot F(p).$$

We obtain the ordinary differential equation

$$p^2 U(x,p) - \frac{1}{LC} U_{xx}(x,p) = 0, \qquad 0 < x < \infty,$$

with the boundary conditions

$$U(0,p) = F(p), \qquad U(\infty,p) = 0$$

(the second of which follows from physical considerations).

Therefore, we have to solve the boundary value problem

$$\begin{cases} U_{xx}(x,p) - p^2 LCU(x,p) = 0, & 0 < x < \infty, \\ U(0,p) = F(p), & U(\infty,p) = 0. \end{cases}$$

The general solution of our ordinary differential equation is

$$U(x,p) = C_1 e^{p\sqrt{LC}\,x} + C_2 e^{-p\sqrt{LC}\,x},$$

where C_1 and C_2 are arbitrary constants. First of all, we must put $C_1 = 0$, because otherwise $U(x,p) \to \infty$ as $x \to \infty$. Hence,

$$U(x,p) = C_2 e^{-p\sqrt{LC}\,x}.$$

Setting here $x = 0$ we get $U(0,p) = C_2$. By assumption, $U(0,p) = F(p)$, and so $C_2 = F(p)$, i.e.,

$$U(x,p) = F(p)e^{-p\sqrt{LC}\,x}.$$

Returning to the original (i.e., applying the inverse Laplace transformation) we get

$$u(x,t) = \begin{cases} E(t - \sqrt{LC}\,x), & \text{if } t > \sqrt{LC}\,x, \\ 0, & \text{if } t > \sqrt{LC}\,x. \end{cases}$$

Now let us consider case (b). Its mathematical formulation is

$$\text{(b)} \quad \begin{cases} u_{xx} = a^2 u_{tt} + 2bu_t + c^2 u, & 0 < x < \infty, \quad 0 < t < \infty, \\ u(0,t) = E(t), & 0 \le t < \infty, \\ u(x,0) = 0, & u_t(x,0) = 0, \quad 0 \le x < \infty, \end{cases}$$

where $a^2 = LC$, $b = (CR + LG)/2$, and $c^2 = RG$, with R and G denoting the resistance and the conductance per unit of length of the conductor.

Applying here the Laplace transformation, we obtain the boundary value problem

$$\begin{cases} U_{xx}(x,p) = a^2 p^2 U(x,p) + 2bpU(x,p) + c^2 U(x,p), & 0 < x < \infty, \\ U(0,p) = F(p), & U(\infty,p) = 0. \end{cases}$$

The general solution of this ordinary differential equation has the form

$$U(x,p) = C_1 e^{a\left(p + \frac{b}{a^2}\right)x} + C_2 e^{-a\left(p + \frac{b}{a^2}\right)x}.$$

The boundary condition $U(\infty,p) = 0$ yields $C_1 = 0$, and so

$$U(x,p) = F(p)e^{-a\left(p + \frac{b}{a^2}\right)x} = F(p)e^{-\frac{b}{a}x}e^{-apx}.$$

This shows that the voltage in the line is

$$u(x,t) = \begin{cases} e^{-\frac{b}{a}x}E(t - ax), & \text{if } t > ax, \\ 0, & \text{if } t < ax. \end{cases}$$

Example 2 [4, no. 839]. Solve the following initial-boundary value problem on the half-line:

$$\begin{cases} u_{tt} = a^2 u_{xx}, & 0 < x < \infty, \quad 0 < t < \infty, \\ u(x,0) = 0, & u_t(x,0) = 0 \quad 0 \le x < \infty, \\ u_x(0,t) - hu(0,t) = \varphi(t), & 0 \le t < \infty. \end{cases}$$

Solution. Again, let us apply the Laplace transformation with respect to the variable t to both sides of the given equation as well as to the boundary and initial conditions. this yields a boundary value problem for an ordinary differential equation:

$$\begin{cases} U_{xx}(x,p) - \dfrac{p^2}{a^2}U(x,p) = 0, & 0 < x < \infty, \\ U_x(0,p) - hU(0,p) = \Phi(p), \end{cases}$$

where $\Phi(p) \doteqdot \varphi(t)$.

As before, the solution of this problem is $U(x,p) = Ce^{-\frac{p}{a}x}$, where C is an arbitrary constant. The boundary condition gives $-\frac{C}{a} - hC = \Phi(p)$, whence $C = -\frac{a\Phi(p)}{ah+p}$ and

$$U(x,p) = -\frac{a\Phi(p)}{ah + p}e^{-\frac{p}{a}x}.$$

Therefore, the original has the form

$$u(x,t) = \begin{cases} 0, & \text{if } x > at, \\ -ae^{h(x-at)}\int_0^{t-x/a} e^{h\tau}\varphi(\tau)\,d\tau, & \text{if } x < at. \end{cases}$$

2.5. The Hankel integral transform method

Suppose given a function $f(\rho)$ $(0 < \rho < \infty)$. If it satisfies certain conditions, then one can define its Hankel transform (of order zero) by the rule

$$\widehat{f}(\lambda) = \int_0^\infty \rho f(\rho) J_0(\lambda\rho)\,d\rho,$$

where J_0 is the Bessel function of first kind of index zero. Then the function $f(\rho)$ is recovered from its image by applying the inverse Hankel transformation:

$$f(\rho) = \int_0^\infty \lambda \widehat{f}(\lambda) J_0(\lambda\rho)\,d\lambda.$$

The function $\widehat{f}(\lambda)$ is called the *Fourier-Bessel-Hankel transform* or *image* of the function $f(\rho)$, which in turn is called the *original*.

Clearly, it is advisable to use the Hankel transformation in the case when the Laplace operator Δu (or the operator $\Delta^2 u$) is written in polar coordinates and the polar radius ranges from 0 to ∞. Let us give some relevant examples.

Example 1 [4, Ch. 4, no. 109]. Solve the boundary value problem

$$\begin{cases} u_{tt} = a^2 \left(\dfrac{\partial^2 u}{\partial \rho^2} + \dfrac{1}{\rho} \dfrac{\partial u}{\partial \rho} \right), & 0 < \rho < \infty, \quad 0 < t < \infty, \\ u(\rho,0) = \dfrac{A}{\sqrt{1 + \rho^2/b^2}}, & u_t(\rho,0) = 0, \qquad 0 \le \rho < \infty, \end{cases}$$

where a^2, A, and b^2 are some constants.

Solution. To derive the equation for the image $\widehat{u}(\lambda,t)$, let us apply the Hankel transformation (of order zero) with respect to the variable ρ to the given equation. We have

$$\widehat{u}(\lambda,t) = \int_0^\infty \rho u(\rho,t) J_0(\lambda\rho)\,d\rho, \quad \text{and also} \quad u(\rho,t) = \int_0^\infty \lambda \widehat{u}(\lambda,t) J_0(\lambda\rho)\,d\lambda.$$

The left-hand side of our equation transforms as

$$u_{tt}(\rho,t) \mapsto \widehat{u}_{tt}(\lambda,t).$$

The right-hand side of the equation transforms as follows (here we use integration by parts):

$$\int_0^\infty \frac{1}{\rho} \frac{\partial}{\partial \rho} \left(\rho \frac{\partial u}{\partial \rho} \right) \rho J_0(\lambda\rho)\,d\rho = \int_0^\infty \frac{\partial}{\partial \rho} \left(\rho \frac{\partial u}{\partial \rho} \right) J_0(\lambda\rho)\,d\rho =$$

$$\left. \left(\rho \frac{\partial u}{\partial \rho} J_0(\lambda\rho) \right) \right|_0^\infty - \int_0^\infty \lambda\rho \frac{\partial u}{\partial \rho} J_0'(\lambda\rho)\,d\rho = -\lambda \int_0^\infty \rho \frac{\partial u}{\partial \rho} J_0'(\lambda\rho)\,d\rho =$$

$$= -\lambda \left. \left(u\rho J_0'(\lambda\rho) \right) \right|_0^\infty + \lambda \int_0^\infty u \frac{\partial u}{\partial \rho} (\rho J_0'(\lambda\rho))\,d\rho =$$

$$\lambda \int_0^\infty (J_0'(\lambda\rho) + J_0''(\lambda\rho)\rho\lambda) u\,d\rho = \int_0^\infty (\lambda^2 \rho J_0''(\lambda\rho) + \lambda J_0'(\lambda\rho)) u\,d\rho,$$

where the integrated terms vanish thanks to the condition $u(\infty,t) = u_\rho(\infty,t) = 0$. By definition, the Bessel function $J_0(x)$ satisfies the equation

$$\lambda^2 J_0''(\lambda\rho) + \frac{\lambda}{\rho} J_0'(\lambda\rho) + \lambda^2 J_0(\lambda\rho) = 0,$$

whence

$$\lambda^2 \rho J_0''(\lambda\rho) + \frac{\lambda}{\rho} J_0'(\lambda\rho) = -\rho\lambda^2 J_0(\lambda\rho).$$

Therefore,

$$\int_0^\infty \frac{1}{\rho}\frac{\partial}{\partial\rho}\left(\rho\frac{\partial u}{\partial\rho}\right)\rho J_0(\lambda\rho)\,d\rho = -\lambda^2\int_0^\infty \rho u J_0(\lambda\rho)\,d\rho = -\lambda^2\widehat{u}(\lambda,t).$$

We see that the given partial differential equation is transformed into the ordinary differential equation

$$\widehat{u}_{tt}(\lambda,t) + a^2\lambda^2\widehat{u}(\lambda,t) = 0, \qquad t > 0.$$

Now let us calculate the Fourier-Bessel-Hankel image of the function $u(\rho,0) = A/\sqrt{1+\rho^2/b^2}$. We have

$$\widehat{u}(\lambda,0) = \int_0^\infty \frac{\rho A}{\sqrt{1+\rho^2/b^2}}\,J_0(\rho\lambda)\,d\rho.$$

It is known that

$$\int_0^\infty e^{-\omega\lambda} J_0(\rho\lambda)\,d\lambda = \frac{1}{\sqrt{\omega^2+\rho^2}},$$

whence

$$\int_0^\infty \frac{e^{-\omega\lambda}}{\lambda}\,\lambda J_0(\rho\lambda)\,d\lambda = \frac{1}{\sqrt{\omega^2+\rho^2}}.$$

Therefore, by the inversion formula,

$$\int_0^\infty \frac{1}{\sqrt{\omega^2+\rho^2}}\,\rho J_0(\rho\lambda)\,d\rho = \frac{e^{-\omega\lambda}}{\lambda},$$

which gives

$$\int_0^\infty \frac{Ab}{\sqrt{b^2+\rho^2}}\,\rho J_0(\rho\lambda)\,d\rho = Ab\,\frac{e^{-b\lambda}}{\lambda}.$$

We thus arrive at the Cauchy problem

$$\begin{cases} \widehat{u}_{tt}(\lambda,t) + a^2\lambda^2\widehat{u}(\lambda,t) = 0, & t > 0, \\ \widehat{u}(\lambda,0) = \dfrac{Ab}{\lambda}e^{-b\lambda}, & \widehat{u}_t(\lambda,0) = 0. \end{cases}$$

The solution of this problem is

$$\widehat{u}(\lambda,t) = \frac{Ab}{\lambda}e^{-b\lambda}\cos(a\lambda t).$$

To obtain the solution of the original problem, it remains to apply the inverse Hankel transformation:

$$u(\rho,t) = \int_0^\infty \lambda\widehat{u}(\lambda,t)J_0(\lambda\rho)\,d\lambda = Ab\int_0^\infty e^{-b\lambda}\cos(a\lambda t)J_0(\lambda\rho)\,d\lambda.$$

Since
$$\cos(a\lambda t) = \operatorname{Re} e^{-ia\lambda t},$$

we get
$$u(\rho, t) = Ab \cdot \operatorname{Re} \int_0^\infty e^{-\lambda(b+iat)} J_0(\lambda\rho) \, d\lambda = Ab \cdot \operatorname{Re} \frac{1}{\sqrt{\rho^2 + (b + iat)^2}} \, .$$

Further,
$$\operatorname{Re} \frac{1}{\sqrt{\rho^2 + (b + iat)^2}} = \operatorname{Re} \frac{1}{b} \cdot \frac{1}{\sqrt{1 + \frac{\rho^2 - a^2 t^2}{b^2} + i\frac{2at}{b}}} = \frac{1}{b} \operatorname{Re} \frac{1}{\sqrt{\alpha + i\beta}},$$

where we denote
$$\alpha = 1 + \frac{\rho^2 - a^2 t^2}{b^2} \qquad \beta = \frac{2at}{b} \, .$$

We have
$$\operatorname{Re} \frac{1}{\sqrt{\alpha + i\beta}} = \operatorname{Re} \frac{1}{\sqrt{\alpha^2 + \beta^2} \left(\cos\frac{\varphi}{2} + i\sin\frac{\varphi}{2} \right)}$$

(here we choose the first branch of $\sqrt{\alpha + i\beta}$, corresponding to $k = 0$ in the expression $\sqrt{\alpha + i\beta} = \sqrt{\alpha^2 + \beta^2} \, e^{i(\varphi + 2\pi k)/2}$, where $k = 0, 1$). Obviously,

$$\operatorname{Re} \frac{1}{\sqrt{\alpha + i\beta}} = \frac{\cos(\varphi/2)}{\sqrt{\alpha^2 + \beta^2}},$$

with
$$\tan\varphi = \frac{\beta}{\alpha}$$

and
$$\cos(\varphi/2) = \sqrt{\frac{1 + \cos\varphi}{2}} = \frac{1}{\sqrt{2}} \sqrt{1 + \cos\varphi} = \frac{1}{\sqrt{2}} \sqrt{1 + \frac{\alpha}{\sqrt{\alpha^2 + \beta^2}}} \, .$$

Therefore,
$$\operatorname{Re} \frac{1}{\sqrt{\alpha + i\beta}} = \frac{1}{\sqrt{2}} \sqrt{\frac{1}{\sqrt{\alpha^2 + \beta^2}} + \frac{\alpha}{\alpha^2 + \beta^2}} \, .$$

Thus, the solution of our problem is

$$u(\rho, t) = \frac{A}{\sqrt{2}} \left[\frac{1}{\sqrt{\left(1 + \frac{\rho^2 - a^2 t^2}{b^2}\right)^2 + \left(\frac{2at}{b}\right)^2}} + \frac{1 + \frac{\rho^2 - a^2 t^2}{b^2}}{\left(1 + \frac{\rho^2 - a^2 t^2}{b^2}\right)^2 + \left(\frac{2at}{b}\right)^2} \right]^{1/2} .$$

Example 2 [4, Ch. VI, no. 110]. Find the radially symmetric transversal oscillations of an infinite plate, assuming that its initial position is given (and depends only on the radius) and the initial velocity is set equal to zero.

Solution. It is known that the transversal oscillations of an infinite plate are described by the equation $u_{tt} + b^2 \Delta^2 u = 0$ with prescribed initial conditions. In the present case it is convenient to pass to polar coordinates and write the mathematical formulation of the problem in the form

$$
\begin{cases}
u_{tt} + b^2 \left(\dfrac{\partial^2 u}{\partial \rho^2} + \dfrac{1}{\rho} \dfrac{\partial}{\partial \rho} \right)^2 u = 0, & 0 < \rho < \infty, \quad 0 < t < \infty, \\
u(\rho, 0) = f(\rho), \quad u_t(\rho, 0) = 0, & 0 \le \rho < \infty, \\
u(\infty, t) = u_\rho(\infty, t) = u_{\rho\rho}(\infty, t) = u_{\rho\rho\rho}(\infty, t) = 0, & 0 \le t < \infty.
\end{cases}
$$

Let us calculate the Hankel transform of the function

$$
\left(\frac{\partial^2 u}{\partial \rho^2} + \frac{1}{\rho} \frac{\partial}{\partial \rho} \right)^2 u = \frac{1}{\rho} \frac{\partial}{\partial \rho} \left(\rho \frac{\partial}{\partial \rho} \left(\frac{1}{\rho} \frac{\partial}{\partial \rho} \left(\rho \frac{\partial u}{\partial \rho} \right) \right) \right).
$$

We have (observing that the integrated terms vanish)

$$
\int_0^\infty \frac{1}{\rho} \frac{\partial}{\partial \rho} \left(\rho \frac{\partial}{\partial \rho} \left(\frac{1}{\rho} \frac{\partial}{\partial \rho} \left(\rho \frac{\partial u}{\partial \rho} \right) \right) \right) \rho J_0(\lambda\rho) \, d\rho =
$$

$$
= \int_0^\infty \frac{\partial}{\partial \rho} \left(\rho \frac{\partial}{\partial \rho} \left(\frac{1}{\rho} \frac{\partial}{\partial \rho} \left(\rho \frac{\partial u}{\partial \rho} \right) \right) \right) J_0(\lambda\rho) \, d\rho =
$$

$$
= -\int_0^\infty \rho \frac{\partial}{\partial \rho} \left(\frac{1}{\rho} \frac{\partial}{\partial \rho} \left(\rho \frac{\partial u}{\partial \rho} \right) \right) \frac{\partial}{\partial \rho} (J_0(\lambda\rho)) \, d\rho =
$$

$$
= \int_0^\infty \frac{1}{\rho} \frac{\partial}{\partial \rho} \left(\rho \frac{\partial u}{\partial \rho} \right) \cdot \frac{\partial}{\partial \rho} \left(\rho \frac{\partial}{\partial \rho} (J_0(\lambda\rho)) \right) \, d\rho =
$$

$$
= -\int_0^\infty \rho \frac{\partial u}{\partial \rho} \frac{\partial}{\partial \rho} \left(\frac{1}{\rho} \frac{\partial}{\partial \rho} \left(\rho \frac{\partial}{\partial \rho} (J_0(\lambda\rho)) \right) \right) \, d\rho =
$$

$$
= \int_0^\infty u \frac{\partial}{\partial \rho} \left(\rho \frac{\partial}{\partial \rho} \left(\frac{1}{\rho} \frac{\partial}{\partial \rho} \left(\rho \frac{\partial}{\partial \rho} (J_0(\lambda\rho)) \right) \right) \right) \, d\rho.
$$

But

$$
\frac{\partial}{\partial \rho} \left(\rho \frac{\partial}{\partial \rho} \left(\frac{1}{\rho} \frac{\partial}{\partial \rho} \left(\rho \frac{\partial}{\partial \rho} (J_0(\lambda\rho)) \right) \right) \right) =
$$

$$
= \frac{\partial}{\partial \rho} \left(\rho \frac{\partial}{\partial \rho} \left(\frac{1}{\rho} \frac{\partial}{\partial \rho} (\lambda\rho J_0'(\lambda\rho)) \right) \right) =
$$

$$
= \lambda \frac{\partial}{\partial \rho} \left(\rho \frac{\partial}{\partial \rho} \left(\frac{1}{\rho} (J_0'(\lambda\rho) + \lambda\rho J_0''(\lambda\rho)) \right) \right) =
$$

$$
= \lambda \frac{\partial}{\partial \rho} \left(\rho \frac{\partial}{\partial \rho} \left(\frac{1}{\rho} J_0'(\lambda\rho) + \lambda J_0''(\lambda\rho) \right) \right).
$$

Next, from the Bessel equation

$$\lambda^2 \rho^2 J_0''(\lambda\rho) + \lambda\rho J_0'(\lambda\rho) + \lambda^2 \rho^2 J_0(\lambda\rho) = 0$$

it follows that

$$\frac{1}{\rho} J_0'(\lambda\rho) + \lambda J_0''(\lambda\rho) = -\lambda J_0(\lambda\rho).$$

Therefore,

$$= \lambda \frac{\partial}{\partial\rho}\left(\rho\frac{\partial}{\partial\rho}\left(\frac{1}{\rho}J_0'(\lambda\rho) + \lambda J_0''(\lambda\rho)\right)\right) =$$

$$= -\lambda^3 \frac{\partial}{\partial\rho}\left(\rho J_0'(\lambda\rho)\right) = -\lambda^3 \left[J_0'(\lambda\rho) + \lambda\rho J_0''(\lambda\rho)\right] = \lambda^4 \rho J_0(\lambda\rho).$$

We conclude that

$$\int_0^\infty \left(\frac{\partial^2}{\partial\rho^2} + \frac{1}{\rho}\frac{\partial}{\partial\rho}\right)^2 u\rho J_0(\lambda\rho)\,d\rho = \lambda^4 \int_0^\infty u(\rho,t)\rho J_0(\lambda\rho)\,d\rho = \lambda^4 \widehat{u}(\lambda,t),$$

and so our original problems is transformed into the following Cauchy problem for an ordinary differential equation:

$$\begin{cases} \widehat{u}_{tt} + b^2\lambda^4\widehat{u}(\lambda,t) = 0, & t > 0, \\ \widehat{u}(\lambda,0) = \widehat{f}(\lambda), & \widehat{u}_t(\lambda,0) = 0, \end{cases}$$

where $\widehat{f}(\lambda) = \int_0^\infty f(\rho)\rho J_0(\lambda\rho)\,d\rho$.

Clearly, the solution of the latter is given by

$$\widehat{u}(\lambda,t) = \widehat{f}(\lambda)\cos(b\lambda^2 t).$$

The solution of the original problem is recovered using the inversion formula:

$$u(\rho,t) = \int_0^\infty \lambda\widehat{f}(\lambda)\cos(b\lambda^2 t)J_0(\lambda\rho)\,d\lambda =$$

$$= \int_0^\infty \lambda\cos(b\lambda^2 t)J_0(\lambda\rho)\left(\int_0^\infty \mu f(\mu)J_0(\lambda\mu)\,d\mu\right)d\lambda =$$

$$= \int_0^\infty \mu f(\mu)\left(\int_0^\infty \lambda\cos(b\lambda^2 t)J_0(\lambda\rho)J_0(\lambda\mu)\,d\lambda\right)d\mu.$$

The integral inside parentheses is further transformed as follows:

$$\int_0^\infty \lambda\cos(b\lambda^2 t)J_0(\lambda\rho)J_0(\lambda\mu)\,d\lambda = \mathrm{Re}\int_0^\infty \lambda e^{ibt\lambda^2}J_0(\lambda\rho)J_0(\lambda\mu)\,d\lambda =$$

$$= \text{Re} \, \frac{1}{-2ibt} \, e^{(\rho^2 + \mu^2)/(4ibt)} J_0 \left(\frac{\rho\mu}{-2ibt} \right) =$$

$$= \text{Re} \frac{1}{2bt} \left(i \cos \frac{\rho^2 + \mu^2}{4bt} + \sin \frac{\rho^2 + \mu^2}{4bt} \right) I_0 \left(\frac{\rho\mu}{2bt} \right) =$$

$$= \frac{1}{2bt} \sin \frac{\rho^2 + \mu^2}{4bt} I_0 \left(\frac{\rho\mu}{2bt} \right).$$

Thus, the solution has the form

$$u(\rho, t) = \frac{1}{2bt} \int_0^\infty \mu f(\mu) \sin \frac{\rho^2 + \mu^2}{4bt} I_0 \left(\frac{\rho\mu}{2bt} \right) d\mu.$$

2.6. The method of standing waves. Oscillations of a bounded string

We already know how to solve the wave equation $u_{tt} = a^2 u_{xx}$ on the whole line $-\infty < x < \infty$ (Cauchy problem). The d'Alembert formula expresses the solution as a sum of two travelling waves that propagate in opposite directions. If we now consider the same equation on a bounded interval $0 < x < l$, then travelling waves no longer exist, since they interact with the boundary. Instead of them a different kind of waves arise, called *standing* or *stationary waves*.

For example, let us examine what happens if a string fastened at the points $x = 0$ and $x = l$ is set into motion. To answer this question we must solve the following mixed hyperbolic problem:

$$\begin{cases} u_{tt} = a^2 u_{xx}, & 0 < x < l, \quad 0 < t < \infty, & (2.9) \\ u(x,0) = f(x), & u_t(x,0) = g(x), & 0 \le x \le l, & (2.10) \\ u(0,t) = 0, & u(l,t) = 0, & 0 \le t < \infty; & (2.11) \end{cases}$$

here (2.9) is the given partial differential equation, (2.10) are initial conditions, and (2.11) are boundary conditions. It turns out that in this case travelling waves are reflected at boundaries in such a way that the resulting oscillations do not travel, but preserve their shape in a fixed position, i.e., they are standing waves. For example, the solution $u(x,t) = \sin x \sin(at)$ of the equation $u_{tt} = a^2 u_{xx}$ is a standing wave.

Problem (2.9)–(2.11) will be solved by the Fourier method (the method of separation of variables). The essential steps of this method are treated in sufficient detail in a number of textbooks (see, e.g., [1], [5], [22]). We will recall here only the idea of the method. Assuming that the solution $u(x,t) = X(x)T(t)$ satisfies only the boundary conditions (2.11), one passes from the partial differential equation $u_{tt} = a^2 u_{xx}$ to two ordinary differential equations: $X'' + \lambda X = 0$ and $T'' + \lambda a^2 T = 0$, where $\lambda > 0$ is the separation constant. Combining the solutions of these equations, we obtain particular solutions of problem (2.9)–(2.11) of the form $X_n(t)T_n(t)$, where $n = 1, 2, \ldots$

By the superposition principle, the general solution of problem (2.9)–(2.11) is given by a series $u(x,t) = \sum_{n=1}^{\infty} C_n X_n(x) T_n(t)$, where C_n are arbitrary constants. Using the initial conditions, one can determine the coefficients C_n. The block diagram of the described process is shown in Figure 2.11.

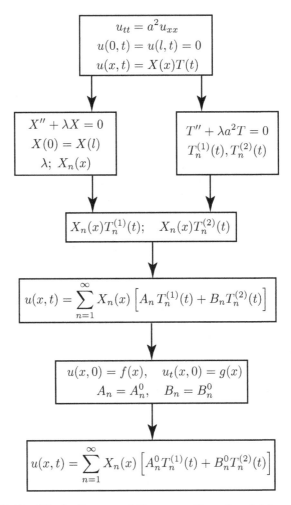

FIGURE 2.11. Block diagram of the separation of variables

Thus, the solution of problem (2.9)–(2.11) is given by the series

$$u(x,t) = \sum_{n=1}^{\infty} \left[a_n \cos \left(\frac{n\pi a}{l} t \right) + b_n \sin \left(\frac{n\pi a}{l} t \right) \right] \sin \left(\frac{n\pi}{l} x \right). \tag{2.12}$$

Setting here first $t = 0$ we get

$$u(x, 0) = \sum_{n=1}^{\infty} a_n \sin\left(\frac{n\pi}{l} x\right).$$

But $u(x, 0) = f(x)$, and so

$$f(x) = \sum_{n=1}^{\infty} a_n \sin\left(\frac{n\pi}{l} x\right).$$

We obtained the Fourier sine series expansion of the function $f(x)$ in the interval $(0, l)$. As is known, its coefficients are computed by the formula

$$a_n = \frac{2}{l} \int_0^l f(x) u(x, 0) \sin\left(\frac{n\pi}{l} x\right) dx.$$

Next, differentiating the series (2.12) with respect to the variable t we get

$$u_t(x, t) = \sum_{n=1}^{\infty} \left[-a_n \frac{n\pi a}{l} \sin\left(\frac{n\pi a}{l} t\right) + b_n \frac{n\pi a}{l} \cos\left(\frac{n\pi a}{l} t\right) \right] \sin\left(\frac{n\pi}{l} x\right). \quad (2.13)$$

Setting here $t = 0$ we obtain

$$u_t(x, 0) = \sum_{n=1}^{\infty} b_n \frac{n\pi a}{l} \sin\left(\frac{n\pi}{l} x\right).$$

But $u_t(x, 0) = g(x)$, and so

$$g(x) = \sum_{n=1}^{\infty} b_n \frac{n\pi a}{l} \sin\left(\frac{n\pi}{l} x\right).$$

It follows that

$$b_n \frac{n\pi a}{l} = \frac{2}{l} \int_0^l g(x) \sin\left(\frac{n\pi}{l} x\right) dx,$$

i.e.,

$$b_n = \frac{2}{n\pi a} \int_0^l g(x) \sin\left(\frac{n\pi}{l} x\right) dx.$$

Thus, the coefficients a_n and b_n are determined by the initial conditions.

Remark 1. When a nonstationary problem is solved by means of the method of separation of variables, it is essential that the boundary conditions be homogeneous. In other words, if the boundary conditions are not homogeneous, then on has first to reduce the problem to one with homogeneous (null) boundary conditions.

Remark 2. If we wish to solve the nonhomogeneous equation $u_{tt} = a^2 u_{xx} + f(x,t)$ with homogeneous boundary and initial conditions, then we need first to find the eigenfunctions of the corresponding homogeneous problem ($u_{tt} = a^2 u_{xx}$ plus homogeneous boundary conditions; in the present case, for instance, the eigenfunctions are $X_n(x) = \sin\left(\frac{n\pi}{l} x\right)$, $n = 1, 2, \ldots$), and then seek the solution of the given problem in the form $u(x,t) = \sum_{n=1}^{\infty} u_n(t) \sin\left(\frac{n\pi}{l} x\right)$, where the functions $u_n(t)$ are subject to determination. This approach to solving a problem is called *the method of expansion of the solution in eigenfunctions*.

Remark 3. Suppose we wish to solve the following hyperbolic problem:

$$\begin{cases} u_{tt} = a^2 u_{xx} + f(x,t), & 0 < x < l, \quad 0 < t < \infty, \\ u(0,t) = 0, \quad u(l,t) = 0, & 0 \le t < \infty, \\ u(x,0) = \varphi(x), \quad u_t(x,0) = \psi(x), & 0 \le x \le l. \end{cases}$$

According to the superposition principle, the solution of this problem can be represented as the sum of the solutions of two hyperbolic problems:

$$(\text{I}) \quad \begin{cases} u_{tt} = a^2 u_{xx} + f(x,t), \\ u(0,t) = 0, \quad u(l,t) = 0, \\ u(x,0) = 0, \quad u_t(x,0) = 0, \end{cases}$$

and

$$(\text{II}) \quad \begin{cases} u_{tt} = a^2 u_{xx}, \\ u(0,t) = 0, \quad u(l,t) = 0, \\ u(x,0) = \varphi(x), \quad u_t(x,0) = \psi(x). \end{cases}$$

2.7. Some examples of mixed problems for the equation of oscillations of a string

Example 1. The mixed problem

$$\begin{cases} u_{tt} = a^2 u_{xx}, & 0 < x < \pi, \quad 0 < t < \infty, \\ u(x,0) = \sin(3x), \quad u_t(x,0) = 0, & 0 \le x \le \pi, \\ u(0,t) = 0, \quad u(\pi,0) = 0, & 0 \le t < \infty \end{cases}$$

admits the solution $u(x,t) = \sin(3x)\cos(3at)$, as one can readily verify directly.

Example 2. Consider the problem

$$\begin{cases} u_{tt} = a^2 u_{xx} + t \sin x, & 0 < x < \pi, \quad 0 < t < \infty, \\ u(x,0) = 0, \quad u_t(x,0) = 0, & 0 \le x \le \pi, \\ u(0,t) = 0, \quad u(\pi,0) = 0, & 0 \le t < \infty. \end{cases}$$

Let us seek its solution in the form $u(x,t) = \sin x \cdot v(t)$, where the function $v(t)$ need to be determined. Substituting this expression of $u(x,t)$ in our non-homogeneous equation we obtain the following ordinary differential equation for $v(t)$:

$$v'' \sin x = -a^2 \sin x \cdot v + t \sin x,$$

whence

$$v'' + a^2 v = t.$$

From the boundary conditions it follows that $v(0) = v'(0) = 0$. Therefore, $v(t)$ is a solution of the Cauchy problem

$$\begin{cases} v'' + a^2 v = t, & 0 \le t < \infty, \\ v(0) = v'(0) = 0. \end{cases}$$

The general solution of the differential equation $v'' + a^2 v = t$ is given by $v(t) = C_1 \cos(at) + C_2 \sin(at) + t/a^2$. The initial conditions readily yield

$$C_1 = 0, \quad c_2 = -\frac{1}{a^3}.$$

Hence,

$$v(t) = \frac{1}{a^2}\left[t - \frac{1}{a}\sin(at)\right],$$

and consequently the sought solution of the mixed problem is

$$u(x,t) = \frac{\sin x}{a^2}\left[t - \frac{1}{a}\sin(at)\right].$$

Example 3 [6, Ch. VI, no 20.14(5)]. Solve the following mixed problem:

$$\begin{cases} u_{tt} = u_{xx} + u, & 0 < x < 2, \quad t > 0, \\ u|_{x=0} = 2t, \quad u|_{x=2} = 0, & t \ge 0, \\ u|_{t=0} = 0, \quad u_t|_{t=0} = 0, & 0 \le x \le 2. \end{cases}$$

Solution. We will seek the function $u(x,t)$ in the form of a sum $u(x,t) = v(x,t) + w(x,t)$, where $w(x,t) = t(2-x)$ is a function satisfying the given boundary conditions. Then the function $v(x,t)$ satisfies the null boundary conditions $v(0,t) = v(2,t) = 0$, the equation $v_{tt} = v_{xx} + v + t(2-x)$ and the initial conditions $v(x,0) = 0$, $v_t(x,0) = x - 2$.

Thus, our task is now to find the solution $v(x,t)$ of the problem

$$\begin{cases} v_{tt} = v_{xx} + v + t(2-x), & 0 < x < 2, \quad t > 0, \\ v(0,t) = v(2,t) = 0, & t \ge 0, \\ v(x,0) = 0, \quad v_t(x,0) = x - 2, & 0 \le x \le 2. \end{cases}$$

As a first step let us solve the following problem: find a solution of the equation $v_{tt} = v_{xx} + v$, not identically equal to zero, which satisfies the homogeneous boundary conditions $v(0,t) = v(2,t) = 0$ and is representable in the form $v(x,t) = X(x)T(t)$. We get the equation

$$T''X = X''T + XT,$$

or, upon dividing both sides by XT,

$$\frac{T''}{T} - 1 = \frac{X''}{X}.$$

The left- [resp., right-] hand side of this equation depends only on t [resp., x]. Since t and x are independent variables,

$$\frac{T''}{T} - 1 = \frac{X''}{X} = -\lambda,$$

where λ is the separation constant. This yields two ordinary differential equations:

$$X'' + \lambda X = 0, \tag{2.14}$$

and

$$T'' + (\lambda - 1)T = 0. \tag{2.15}$$

From the boundary conditions $v(0,t) = v(2,t) = 0$ and the expression of $v(t,x)$ it follows that $X(0)T(t) = 0$ and $X(2)T(t) = 0$, whence $X(0) = X(2) = 0$. Thus, we arrived at the Sturm-Liouville problem

$$\begin{cases} X'' + \lambda X = 0, & 0 < x < 2, \\ X(0) = X(2) = 0. \end{cases}$$

The eigenvalues and corresponding eigenfunctions of this problem are $\lambda_n = \left(\frac{n\pi}{2}\right)^2$ and $X_n(x) = \sin\left(\frac{n\pi}{2}x\right)$, $n = 1,2,\dots$. Next, from equation (2.15) it follows that

$$T_n(t) = A_n \cos(\mu_n t) + B_n \sin(\mu_n t),$$

where A_n and B_n are arbitrary constants and $\mu_n = \sqrt{\left(\frac{n\pi}{2}\right)^2 - 1}$, $n = 1,2,\dots$ Clearly, the functions

$$v_n(x,t) = [A_n \cos(\mu_n t) + B_n \sin(\mu_n t)] \sin\left(\frac{n\pi}{2}x\right)$$

are particular solutions of the equation $v_{tt} = v_{xx} + v$ that satisfy null boundary conditions.

Now let us solve the following two problems

$$\text{(I)}\quad \begin{cases} v_{tt} = v_{xx} + v, & 0 < x < 2, \quad t > 0, \\ v(x,0) = 0, & v_t(x,0) = x - 2, \quad 0 \le x \le 2, \\ v(0,t) = 0, & v(2,t) = 0, \quad t \ge 0, \end{cases} \qquad (2.16)$$

and

$$\text{(II)}\quad \begin{cases} v_{tt} = v_{xx} + v + t(2 - x), & 0 < x < 2, \quad t > 0, \\ v(x,0) = 0, & v_t(x,0) = 0, \quad 0 \le x \le 2, \\ v(0,t) = 0, & v(2,t) = 0, \quad t \ge 0. \end{cases} \qquad (2.17)$$

The solution of problem (2.16) is given by the series

$$P(x,t) = \sum_{n=1}^{\infty} [A_n \cos(\mu_n t) + B_n \sin(\mu_n t)] \sin\left(\frac{n\pi}{2} x\right),$$

whose coefficients A_n and B_n are given by the formulas

$$A_n = \int_0^2 v(x,0) \sin\left(\frac{n\pi}{2} x\right) dx = 0,$$

and

$$B_n = \frac{1}{\mu_n} \int_0^2 v_t(x,0) \sin\left(\frac{n\pi}{2} x\right) dx = \frac{1}{\mu_n} \int_0^2 (x - 2) \sin\left(\frac{n\pi}{2} x\right) dx = -\frac{4}{n\pi\mu_n}.$$

Therefore,

$$P(x,t) = \sum_{n=1}^{\infty} \left(-\frac{4}{n\pi\mu_n}\right) \sin(\mu_n t) \sin\left(\frac{n\pi}{2} x\right).$$

The solution of problem (2.17) will be sought in the form

$$Q(x,t) = \sum_{n=1}^{\infty} q_n(t) \sin\left(\frac{n\pi}{2} x\right)$$

(here we use the method of expansion in eigenfunctions). Substituting this expression in the equation of problem (2.17) and observing that

$$t(2 - x) = \sum_{n=1}^{\infty} \frac{4t}{n\pi} \sin\left(\frac{n\pi}{2} x\right),$$

we obtain

$$\sum_{n=1}^{\infty} q_n''(t) \sin\left(\frac{n\pi}{2} x\right) = -\sum_{n=1}^{\infty} \left(\frac{n\pi}{2}\right)^2 q_n(t) \sin\left(\frac{n\pi}{2} x\right) +$$

$$+ \sum_{n=1}^{\infty} q_n(t) \sin\left(\frac{n\pi}{2} x\right) + \sum_{n=1}^{\infty} \frac{4t}{n\pi} \sin\left(\frac{n\pi}{2} x\right)$$

or

$$\sum_{n=1}^{\infty} \sin\left(\frac{n\pi}{2}x\right)\left[q_n'' + \left(\left(\frac{n\pi}{2}\right)^2 - 1\right)q_n - \frac{4t}{n\pi}\right] = 0.$$

This yields the following Cauchy problem for the determination of the functions $q_n(t)$ $(n = 1, 2, \dots)$:

$$\begin{cases} q_n'' + \mu_n^2 q_n = \dfrac{4t}{n\pi}, & 0 \le t < \infty, \\ q_n(0) = q_n'(0) = 0. \end{cases}$$

But we have already encountered a similar problem, so we will write its solution without further comments:

$$q_n(t) = -\frac{4}{n\pi\mu_n^3}\sin(\mu_n t) + \frac{4t}{n\pi\mu_n^2}.$$

Therefore, the solution of problem (2.17) is given by the series

$$Q(t, x) = \sum_{n=1}^{\infty}\left[-\frac{4}{n\pi\mu_n^3}\sin(\mu_n t) + \frac{4t}{n\pi\mu_n^2}\right]\sin\left(\frac{n\pi}{2}x\right).$$

By the superposition principle, we can write $v(x, t) = P(x, t) + Q(x, t)$, or

$$v(x, t) = \sum_{n=1}^{\infty}\left[\frac{4t}{n\pi\mu_n^2} - \frac{n\pi}{\mu_n^3}\sin(\mu_n t)\right]\sin\left(\frac{n\pi}{2}x\right).$$

We conclude that the solution of our original problem, $u(x, t) = v(x, t) + t(2 - x)$, is

$$u(x, t) = \sum_{n=1}^{\infty}\left[\frac{4t}{n\pi\mu_n^2} - \frac{n\pi}{\mu_n^3}\sin(\mu_n t)\right]\sin\left(\frac{n\pi}{2}x\right) + t(2 - x).$$

Let us show that any mixed problem for the wave equation has a unique solution. To this end we will establish the following assertion.

Proposition. *Suppose the function $u(x, t)$ is twice continuously differentiable in the domain Ω: $0 \le x \le l$, $0 \le t \le T$ and solves the mixed problem*

$$\begin{cases} u_{tt} = a^2 u_{xx}, & 0 < x < l, \quad 0 < t < T, \\ u(0, t) = 0, \quad u(l, t) = 0, & 0 \le t \le T, \\ u(x, 0) = u_t(x, 0) = 0, & 0 \le x \le l, \end{cases}$$

where T is an arbitrarily fixed positive number. Then $u(x, t) \equiv 0$ in Ω.

Proof. Let us multiply the equation $u_{tt} = a^2 u_{xx}$ by u_t and integrate the result over the domain Ω. We obtain

$$\int_0^t \int_0^l u_{tt} u_t \, dx \, dt = a^2 \int_0^t \int_0^l u_{xx} u_t \, dx \, dt.$$

Observing that $u_{tt} u_t = \frac{1}{2} \frac{\partial}{\partial t}(u_t^2)$, we have

$$\int_0^t \int_0^l u_{tt} u_t \, dx \, dt = \int_0^l dx \int_0^t \frac{1}{2} \frac{\partial}{\partial t}(u_t^2) dt =$$

$$= \frac{1}{2} \int_0^l [u_t^2(x,t) - u_t^2(x,0)] \, dx = \frac{1}{2} \int_0^l u_t^2 \, dx.$$

Next, integration by parts gives

$$\int_0^l u_{xx} u_t \, dx = (u_x(x,t) u_t(x,t))\big|_0^l - \int_0^l u_x u_{tx} dx =$$

$$= u_x(l,t) u_t(l,t) - u_x(0,t) u_t(0,t) - \frac{1}{2} \int_0^l \frac{\partial}{\partial t}(u_x^2) dx = -\frac{1}{2} \int_0^l \frac{\partial}{\partial t}(u_x^2) dx,$$

because $u_t(0,t) = u_t(l,t) = 0$ (this is obtained by differentiating the equalities $u(0,t) = u(l,t) = 0$ with respect to t). We deduce that

$$\frac{1}{2} \int_0^l u_t^2 \, dx = -\frac{a^2}{2} \int_0^l dx \int_0^t \frac{\partial}{\partial t}(u_x^2) dt,$$

or

$$\frac{1}{2} \int_0^l u_t^2 \, dx + \frac{a^2}{2} \int_0^l u_x^2 \, dx = 0,$$

because $u_x(x,0) = 0$ (which is shown by differentiating the equality $u(x,0) = 0$ with respect to x).

The quantity

$$E(t) = \int_0^l \left[\frac{1}{2} u_t^2 + \frac{a^2}{2} u_x^2 \right] dx$$

is called the *energy integral*. We have shown that $E(t) = 0$. Further, from this relation it follows that $u_t(x,t) = u_x(x,t) \equiv 0$ in the domain Ω. Therefore, $du(x,t) \equiv 0$ in Ω, and then the Taylor formula implies that $u(x,t) \equiv \text{const}$ in Ω. But $u(x,t)$ is continuous in Ω and $u(x,0) = 0$. We conclude that $u(x,t) \equiv 0$ in the domain Ω.

2.8. The Fourier method. Oscillations of a rectangular membrane

Here *membrane* is understood as a pellicle, i.e., a very thin solid body, uniformly stretched in all directions and which does not resist bending. As it turns out, to describe the oscillations of a membrane one does not need special functions; rather, one can manage with sines and cosines.

Example 1 [4, Ch. VI, no. 20.18]. Solve the following mixed problem:

$$
\begin{cases}
u_{tt} = u_{xx} + u_{yy}, \quad 0 < x, y < \pi, & \text{(2.18)} \\
u|_{x=0} = u|_{x=\pi} = u|_{y=0} = u|_{y=\pi} = 0, \quad t > 0, & \text{(2.19)} \\
u|_{t=0} = 3\sin x \sin(2y), & \\
\quad\quad\quad\quad\quad\quad\quad\quad 0 \le x, y \le \pi. & \text{(2.20)} \\
u_t|_{t=0} = 5\sin(3x)\sin(4y),
\end{cases}
$$

Solution. First let us consider an auxiliary problem: find a function $u(x, y, t)$ satisfying equation (2.18) and the boundary conditions (2.19). We will seek this function in the form $u(x, y, t) = v(x, y)T(t)$. Substituting this expression in equation (2.18) we obtain

$$vT'' = Tv_{xx} + Tv_{yy},$$

which upon dividing both sides by vT yields

$$\frac{T''}{T} = \frac{v_{xx} + v_{yy}}{v}. \tag{2.21}$$

We see that the left-hand side of equation (2.21) depends only on t, whereas the right-hand side depends only on x and y. Since these variables are independent of one another, equality can hold in (2.21) only if

$$\frac{T''}{T} = \frac{v_{xx} + v_{yy}}{v} = -\lambda, \tag{2.22}$$

where λ is a constant. This yields two equations:

the partial differential equation

$$v_{xx} + v_{yy} + \lambda v = 0; \tag{2.23}$$

the ordinary differential equation

$$T'' + \lambda T = 0. \tag{2.24}$$

Now let us remark that from the boundary conditions (2.19) it follows that

$$v(0, y) = 0, \quad v(\pi, y) = 0, \quad v(x, 0) = 0, \quad v(x, \pi) = 0.$$

Thus, for the function $v(x, y)$ we have the following boundary value problem:

$$\begin{cases} v_{xx} + v_{yy} + \lambda v = 0, & 0 < x, y < \pi, \\ v(0, y) = v(\pi, y) = v(x, 0) = v(x, \pi) = 0, & 0 \leq x, y \leq \pi. \end{cases} \tag{2.25, 2.26}$$

Our task is now to find the values of λ for which problem (2.25), (2.26) has a nonzero (nontrivial) solution and find that solution.

To this end we use again the method of separation of variables, setting

$$v(x, y) = X(x)Y(y). \tag{2.27}$$

Substituting this expression in equation (2.25) we get

$$X''Y + XY'' + \lambda XY = 0,$$

whence, upon dividing by XY,

$$\frac{X''}{X} = -\frac{Y''}{Y} - \lambda. \tag{2.28}$$

The left- [resp., right-] hand side of this equation depends only on x [resp., y]. Consequently, equality can hold only when

$$\frac{X''}{X} = -\frac{Y''}{Y} - \lambda = -\mu, \tag{2.29}$$

where μ is a new separation constant. This yields two ordinary differential equations:

$$X'' + \mu X = 0, \tag{2.30}$$

and

$$Y'' + (\lambda - \mu)Y = 0. \tag{2.31}$$

From the boundary conditions (2.26) it follows that

$$X(0) = X(\pi) = 0, \quad Y(0) = Y(\pi) = 0. \tag{2.32}$$

We thus arrive at two Sturm-Liouville problems:

$$\begin{cases} X'' + \mu X = 0, & 0 < x < \pi, \\ X(0) = X(\pi) = 0; \end{cases} \tag{2.33}$$

and

$$\begin{cases} Y'' + \nu Y = 0, & 0 < y < \pi, \nu = \lambda - \mu, \\ Y(0) = Y(\pi) = 0; \end{cases} \tag{2.34}$$

We have already solved such problems, so we know that their solutions have the form

$$X_n(x) = \sin(nx), \quad \mu_n = n^2, \quad n = 1, 2, \ldots,$$

$$Y_m(y) = \sin(my), \quad \nu = m^2, \quad m = 1, 2, \ldots$$

Then to the eigenvalues

$$\lambda_{nm} = \mu_n + \nu_m = n^2 + m^2$$

will correspond the eigenfunctions

$$v_{nm}(x, y) = \sin(nx)\sin(my), \quad n, m = 1, 2, \ldots$$

Now let us return to equation (2.24) and substitute there $\lambda_{nm} = n^2 + m^2$ for λ. We obtain the equation

$$T''_{nm} + \lambda_{nm} T_{nm} = 0.$$

The general solution of this equation is

$$T_{nm}(t) = a_{nm} \cos\left(\sqrt{n^2 + m^2}\, t\right) + b_{nm} \sin\left(\sqrt{n^2 + m^2}\, t\right),$$

where a_{nm} and b_{nm} are arbitrary constants.

Therefore, the solution "atoms" of problem (2.18), (2.19) are the products

$$v_{nm}(x, y) T_{nm}(t).$$

The solution of problem (2.18)–(2.20) is given by the series

$$u(x, y, t) = \sum_{n,m=1}^{\infty} \left[a_{nm} \cos\left(\sqrt{n^2 + m^2}\, t\right) + b_{nm} \sin\left(\sqrt{n^2 + m^2}\, t\right)\right] \sin(nx)\sin(my),$$

$$(2.35)$$

where the coefficients a_{nm} and b_{nm} are subject to determination.

One can verify that the system of functions $\{\sin(nx)\sin(my)\}_{n,m=1}^{\infty}$ is orthogonal and complete in the rectangle $0 \leq x, y \leq \pi$ (orthogonality is obvious because

$$\int_0^\pi \int_0^\pi \sin(n_1 x)\sin(m_1 y)\sin(n_2 x)\sin(m_2 y)\, dx\, dy =$$

$$\int_0^\pi \sin(n_1 x)\sin(n_2 x)\, dx \int_0^\pi \sin(m_1 y)\sin(m_2 y)\, dy = 0$$

whenever $n_1 \neq n_2$ or $m_1 \neq m_2$; the proof of the completeness is more difficult and is omitted).

Now if we put $t = 0$ in the general solution, we obtain

$$u(x, y, 0) = \sum_{n,m=1}^{\infty} a_{nm} \sin(nx) \sin(my).$$

Using the formula for the coefficients of a double Fourier series, we have

$$a_{nm} = \frac{4}{\pi^2} \int_0^{\pi} \int_0^{\pi} u(x, y, 0) \sin(nx) \sin(my) \, dx \, dy.$$

In our case $u(x, y, 0) = 3 \sin x \sin(2y)$. Hence,

$$a_{12} = \frac{4}{\pi^2} \int_0^{\pi} \int_0^{\pi} e \sin^2 x \sin^2(2y) \, dx \, dy = 3, \qquad a_{nm} = 0 \text{ if } n \neq 1, m \neq 2.$$

Further, differentiating the solution (2.35) with respect to t, we obtain

$$u_t(x, y, t) = [-3\sqrt{5}] \sin(\sqrt{5}t) \sin x \sin(2y) +$$

$$+ \sum_{n,m=1}^{\infty} b_{nm} \sqrt{n^2 + m^2} \cos(\sqrt{n^2 + m^2} t) \sin(nx) \sin(my).$$

Putting here $t = 0$, we have

$$u_t(x, y, 0) = \sum_{n,m=1}^{\infty} \sqrt{n^2 + m^2} b_{nm} \sin(nx) \sin(my),$$

or

$$5 \sin(3x) \sin(4y) = \sum_{n,m=1}^{\infty} \sqrt{n^2 + m^2} b_{nm} \sin(nx) \sin(my).$$

It follows that

$$b_{34} = \frac{1}{5} \frac{4}{\pi^2} \int_0^{\pi} \int_0^{\pi} 5 \sin^2(3x) \sin^2(4y) \, dx \, dy = 1, \qquad b_{nm=0} \text{ if } n \neq 3, m \neq 4.$$

Thus, the solution of the problem (2.18)-(2.20) is the function

$$u(x, y, t) = 3 \cos(\sqrt{5}t) \sin x \sin(2y) + \sin(5t) sin(3x) sin(4y).$$

Remark. The solution of the mixed problem

$$\begin{cases} u_{tt} = a^2(u_{xx} + u_{yy}) + f(x, y, t), & 0 < x, y < b, \quad t > 0, \\ u(x, y, 0) = u_t(x, y, 0) = 0, & 0 \leq x, y \leq b, \\ u(0, y, t) = u(b, y, t) = 0, & u(x, 0, t) = u(x, b, t) = 0, \quad t \geq 0, \end{cases}$$

can be obtained by the method of expansion in the eigenfunctions of the corresponding homogeneous problem. Specifically, one seeks the solution in the form

$$u(x, y, t) = \sum_{n,m=1}^{\infty} u_{nm}(t) \sin\left(\frac{n\pi}{b}x\right) \sin\left(\frac{n\pi}{b}y\right),$$

where the functions $u_{nm}(t)$ are found in much the same manner as in the case of the solution of the analogous problem for the nonhomogeneous equation of the oscillations of a string.

Example 2. Solve the following mixed problem (with Neumann boundary conditions)

$$
\begin{cases}
u_{tt} = u_{xx} + u_{yy} + \cos t \cos x \cos y, & 0 < x, y < \pi, \quad t > 0, & (2.36) \\
\left.\dfrac{\partial u}{\partial n}\right|_{\Gamma}, \quad t \geq 0, & (2.37) \\
u(x, y, 0) = u_t(x, y, 0) = 0, \quad 0 \leq x, y \leq \pi, & (2.38)
\end{cases}
$$

where Γ is the boundary of the square $\{(x, y) : 0 \leq x, y \leq \pi\}$.

Solution. As we just indicated, we will seek the solution in the form $u(x, y, t) = v(t) \cos x \cos y$ (because the eigenfunctions of the corresponding homogeneous problem

$$
\begin{cases}
u_{tt} = u_{xx} + u_{yy}, & 0 < x, y < b, \quad t > 0, \\
\left.\dfrac{\partial u}{\partial n}\right|_{\Gamma} = 0, & t \geq 0,
\end{cases}
$$

are the products $\cos(nx)\cos(my)$).
Substituting this expression of the function $u(x, y, t)$ in equation (2.36) we obtain the Cauchy problem

$$
\begin{cases}
v''(t) + 2v(t) = \cos t, & t \geq 0, & (2.39) \\
v(0) = v'(0) = 0. & (2.40)
\end{cases}
$$

The general solution of problem (2.39), (2.40) has the form

$$v(t) = C_1 \cos(\sqrt{2}t) + C_2 \sin(\sqrt{2}t) + \cos t,$$

where C_1 and C_2 are arbitrary constants. Imposing the initial conditions (2.40), we get $C_1 = -1$, $C_2 = 0$. Consequently,

$$v(t) = \cos t - \cos(\sqrt{2}t).$$

Therefore, the solution of the original problem (2.36)–(2.38) is

$$u(x, y, t) = \cos x \cos y[\cos t - \cos(\sqrt{2}t)].$$

Example 3 [18, § 3, no. 106]. A rectangular membrane clamped along its edges at the initial time $t = 0$, is subject to an impact in a neighborhood of its center, so that

$$\lim_{\varepsilon \to 0} \iint_{\Omega_\varepsilon} v_0(x, y) \, dx \, dy = 0,$$

where $v_0(x, y)$ is the initial velocity, $A = \text{const}$, and Ω_ε is a neighborhood of the central point. Determine the resulting free oscillations of the membrane.

Solution. The problem is formulated mathematically as follows:

$$\begin{cases} u_{tt} = a^2(u_{xx} + u_{yy}), & 0 < x < b_1, \quad 0 < y < b_2, \quad t > 0, \\ u|_{x=0} = u|_{x=b_1} = u|_{y=0} = u|_{y=b_2} = 0, & t \geq 0, \\ u|_{t=0} = 0, \quad u_t|_{t=0} = \begin{cases} v_0(x, y) & \text{if } (x, y) \in \Omega_\varepsilon(b_1/2, b_2/2), \\ 0, & \text{if } (x, y) \notin \Omega_\varepsilon(b_1/2, b_2/2), \end{cases} \end{cases}$$

where $\Omega_\varepsilon(b_1/2, b_2/2)$ is the open disc of radius ε centered at the point $(b_1/2, b_2/2)$. Applying the method of separation of variables, we obtain

$$u(x, y, t) = \sum_{n,m=1}^{\infty} [A_{nm} \cos(\lambda_{nm} at) + B_{nm} \sin(\lambda_{nm} at)] \sin\left(\frac{n\pi}{b_1} x\right) \sin\left(\frac{m\pi}{b_2} y\right).$$

Here $\lambda_{nm} = \pi\sqrt{(n/b_1)^2 + (m/b_2)^2}$.

It remains to find the coefficients A_{nm} and B_{nm}. Setting $t = 0$ in the above expression we get

$$u(x, y, 0) = \sum_{n,m=1}^{\infty} A_{nm} \sin\left(\frac{n\pi}{b_1} x\right) \sin\left(\frac{m\pi}{b_2} y\right).$$

Since we assume that $u(x, y, 0) = 0$, it follows that $A_{nm} = 0$, $n, m = 1, 2, \ldots$, i.e.,

$$u(x, y, t) = \sum_{n,m=1}^{\infty} B_{nm} \sin(\lambda_{nm} at) \sin\left(\frac{n\pi}{b_1} x\right) \sin\left(\frac{m\pi}{b_2} y\right).$$

Differentiating this equality with respect to t, we have

$$u_t(x, y, t) = \sum_{n,m=1}^{\infty} B_{nm} \lambda_{nm} a \cos(\lambda_{nm} at) \sin\left(\frac{n\pi}{b_1} x\right) \sin\left(\frac{m\pi}{b_2} y\right),$$

whence

$$u_t(x, y, 0) = \sum_{n,m=1}^{\infty} B_{nm} \lambda_{nm} a \sin\left(\frac{n\pi}{b_1} x\right) \sin\left(\frac{m\pi}{b_2} y\right).$$

Therefore,

$$B_{nm} = \frac{4}{b_1 b_2 a \lambda_{nm}} \int_0^{b_1} \int_0^{b_2} v_0(x,y) \sin\left(\frac{n\pi}{b_1} x\right) \sin\left(\frac{m\pi}{b_2} y\right) dx\, dy =$$

$$= \frac{4}{b_1 b_2 a \lambda_{nm}} \iint_{\Omega_\varepsilon} v_0(x,y) \sin\left(\frac{n\pi}{b_1} x\right) \sin\left(\frac{m\pi}{b_2} y\right) dx\, dy =$$

$$= \frac{4}{b_1 b_2 a \lambda_{nm}} \sin\left(\frac{n\pi}{b_1} \overline{x}\right) \sin\left(\frac{m\pi}{b_2} \overline{y}\right) \iint_{\Omega_\varepsilon} dx\, dy$$

where $(\overline{x}, \overline{y}) \in \Omega_\varepsilon$ (by the mean value theorem).

Passing to the limit $\varepsilon \to 0$ in this equality, we obtain

$$B_{nm} = \frac{4A}{b_2 b_2 a \lambda_{nm}} \sin\left(\frac{n\pi}{b_1}\right) \sin\left(\frac{m\pi}{b_2}\right), \qquad n, m = 1, 2, \ldots$$

Therefore, the solution of our problem is the function

$$u(x,y,t) = \frac{4A}{b_2 b_2 a} \sum_{n,m=1}^{\infty} \frac{\sin\left(\frac{n\pi}{b_1}\right) \sin\left(\frac{m\pi}{b_2}\right)}{\lambda_{n,m}} \times$$

$$\times \sin(\lambda_{nm} a t) \sin\left(\frac{n\pi}{b_1} x\right) \sin\left(\frac{m\pi}{b_2} y\right).$$

Remark. The problem considered in Example 3 can also be solved using the notion of δ-function.

2.9. The Fourier method. Oscillations of a circular membrane

To describe the oscillations of a circular membrane one needs to resort to special functions, namely, Bessel functions.

Example 1 [18, §3, no. 104]. A homogeneous circular membrane of radius R, clamped along its contour, is in a state of equilibrium under a tension T. At time $t = 0$ the uniformly distributed load $f = P_0 \sin(\omega t)$ is applied to the surface of the membrane. Find the radial oscillations of the membrane.

Solution. The problem is mathematically formulated as follows:

$$\begin{cases} \dfrac{1}{a^2} u_{tt} = \dfrac{1}{\rho} \dfrac{\partial}{\partial \rho} \left(\rho \dfrac{\partial u}{\partial \rho}\right) + \dfrac{P_0}{T} \sin(\omega t), & 0 < \rho < R, \quad t > 0, \\ |u(0,t)| < \infty, \quad u(R,t) = 0, & t \geq 0, \\ u(\rho, 0) = 0, \quad u_t(\rho, 0) = 0, & 0 \leq \rho \leq R. \end{cases} \qquad (2.41)$$

By the superposition principle, the solution of this problem can be represented as a sum $u(\rho, t) = w(\rho, t) + v(\rho, t)$, where $w(\rho, t)$ is the solution of the boundary value problem

$$\begin{cases} \dfrac{1}{a^2} w_{tt} = \dfrac{1}{\rho} \dfrac{\partial}{\partial \rho} \left(\rho \dfrac{\partial w}{\partial \rho} \right) + \dfrac{P_0}{T} \sin(\omega t), & 0 < \rho < R, \quad t > 0, \\ |w(0, t)| < \infty, \quad u(R, t) = 0, & t \geq 0, \end{cases} \tag{2.42}$$

and $v(\rho, t)$ is the solution of the problem

$$\begin{cases} \dfrac{1}{a^2} v_{tt} = \dfrac{1}{\rho} \dfrac{\partial}{\partial \rho} \left(\rho \dfrac{\partial v}{\partial \rho} \right), & 0 < \rho < R, \quad t > 0, \\ |v(0, t)| < \infty, \quad v(R, t) = 0, & t \geq 0, \\ v(\rho, 0) = -w(\rho, 0), \quad v_t(\rho, 0) = -w_t(\rho, 0), & 0 \leq \rho \leq R. \end{cases} \tag{2.43}$$

Obviously, we must first find $w(\rho, t)$. To this end we will seek $w(\rho, t)$ in the form $w(\rho, t) = A(\rho) \sin(\omega t)$, where the function $A(\rho)$ needs to be determined. Substituting this expression into the equation of problem (2.42), we obtain

$$-\frac{1}{a^2} \omega^2 A \sin(\omega t) = \frac{1}{\rho} \frac{d}{d\rho} \left(\rho \frac{dA}{d\rho} \right) \sin(\omega t) + \frac{P_0}{T} \sin(\omega t).$$

Hence, after division by $\sin(\omega t)$, we arrive at the following boundary value problem for an ordinary differential equation:

$$\begin{cases} \rho^2 A'' + \rho A' + \dfrac{\rho^2 \omega^2}{a^2} A = -\dfrac{P_0}{T} \rho^2, & 0 < \rho < R, \\ |A(0)| < \infty, \quad A(R) = 0. \end{cases} \tag{2.44}$$

The solution of problem (2.44) is $A(\rho) = A_1(\rho) + A_2(\rho)$, where $A_1(\rho)$ is the solution of the problem

$$\begin{cases} \rho^2 A_1'' + \rho A_1' + \dfrac{\rho^2 \omega^2}{a^2} A_1 = 0, & 0 < \rho < R, \\ |A_1(0)| < \infty, \quad A_1(R) = 0, \end{cases} \tag{2.45}$$

and $A_2(\rho)$ is a particular solution of the differential equation

$$\rho^2 A_2'' + \rho A_2' + \frac{\rho^2 \omega^2}{a^2} A_2 = -\frac{P_0}{T} \rho^2, \qquad 0 < \rho < R. \tag{2.46}$$

It is readily seen that equation (2.45) reduces via the substitution $x = \omega \rho / a$ to the Bessel equation of order zero, and in view of the boundary condition $|A_1(0)| < \infty$, the solution of problem (2.45) is of the form $A_1(\rho) = C J_0(\omega \rho / a)$, where $J_0(x)$

is the Bessel function of the first kind and of order zero. Further, a solution of equation (2.46) is provided by the constant function

$$A_2(\rho) = -\frac{P_0}{T}\frac{a^2}{\omega}.$$

Therefore,

$$A(\rho) = C J_0\left(\frac{\omega\rho}{a}\right) - \frac{P_0}{T}\frac{a^2}{\omega}.$$

Imposing the boundary condition $A(R) = 0$ we have

$$0 = C J_0\left(\frac{\omega R}{a}\right) - \frac{P_0}{T}\frac{a^2}{\omega},$$

whence

$$C = \frac{P_0}{T}\frac{a^2}{\omega^2}\frac{1}{J_0\left(\frac{\omega R}{a}\right)}.$$

We conclude that

$$A(\rho) = \frac{P_0}{T}\frac{a^2}{\omega^2}\left[\frac{J_0\left(\frac{\omega\rho}{a}\right)}{J_0\left(\frac{\omega R}{a}\right)} - 1\right],$$

and so

$$w(\rho,t) = \frac{P_0}{T}\frac{a^2}{\omega^2}\left[\frac{J_0\left(\frac{\omega\rho}{a}\right)}{J_0\left(\frac{\omega R}{a}\right)} - 1\right]\sin(\omega t).$$

Now let us find the function $v(\rho,t)$, solving problem (2.43). We note that the initial conditions are

$$v(\rho,0) = 0, \qquad v_t(\rho,0) = -\frac{P_0}{T}\frac{a^2}{\omega^2}\left[\frac{J_0\left(\frac{\omega\rho}{a}\right)}{J_0\left(\frac{\omega R}{a}\right)} - 1\right].$$

Writing $v(\rho,t) = B(\rho)T(t)$, substituting this expression in the equation of problem (2.43) and separating the variables we obtain

$$\frac{T''}{a^2 T} = -\frac{1}{B}\frac{1}{\rho}\frac{d}{d\rho}(\rho B') = -\lambda.$$

This yields two ordinary differential equations:

$$\rho^2 B'' + \rho B' + \lambda\rho^2 B = 0, \tag{2.47}$$

and

$$T'' + \lambda a^2 T = 0. \tag{2.48}$$

The change of variables $x = \sqrt{\lambda}\rho$ reduces equation (2.47) to the Bessel equation of order zero. Therefore, the solution of equation (2.47) with the boundary conditions $|B(0)| < \infty$, $B(R) = 0$ is

$$B(\rho) = C J_0\left(\frac{\mu_n}{R} \rho\right) \quad C = \text{const},$$

for values $\lambda_n = \left(\frac{\mu_n}{R}\right)^2$, $n = 1, 2, \ldots$

Next, from equation (2.48) we obtain the sequence of solutions

$$T_n(t) = C_1 \cos\left(a\,\frac{\mu_n}{R}\,t\right) + C_2 \sin\left(a\,\frac{\mu_n}{R}\,t\right),$$

where C_1 and C_2 are arbitrary constants.

The solution of problem (2.43) is given by a series

$$v(\rho, t) = \sum_{n=1}^{\infty} \left[b_n \cos\left(a\,\frac{\mu_n}{R}\,t\right) + c_n \sin\left(a\,\frac{\mu_n}{R}\,t\right) \right] J_0\left(\frac{\mu_n}{R}\,\rho\right).$$

Let us choose the constants b_n and c_n so that the boundary conditions

$$v(\rho, 0) = 0, \qquad v_t(\rho, 0) = -\frac{P_0 a^2}{\omega T}\left[\frac{J_0\left(\frac{\omega}{a}\rho\right)}{J_0\left(\frac{\omega}{a}R\right)} - 1\right]$$

will be satisfied. Clearly, $b_n = 0$, since

$$\sum_{n=1}^{\infty} b_n J_0\left(\frac{\mu_n}{R}\rho\right) = 0.$$

Further, differentiating the function $v(\rho, t)$ with respect to t we have

$$v_t(\rho, t) = \sum_{n=1}^{\infty} c_n a\,\frac{\mu_n}{R} \cos\left(a\,\frac{\mu_n}{R}\,t\right) J_0\left(\frac{\mu_n}{R}\rho\right).$$

Setting here $t = 0$ we obtain

$$-\frac{P_0 a^2}{\omega T}\left[\frac{J_0\left(\frac{\omega}{a}\rho\right)}{J_0\left(\frac{\omega}{a}R\right)} - 1\right] = \sum_{n=1}^{\infty} c_n a\,\frac{\mu_n}{R} \cos\left(a\,\frac{\mu_n}{R}\,t\right) J_0\left(\frac{\mu_n}{R}\rho\right).$$

To calculate the coefficients c_n, let us multiply both sides of this equality by $\rho J_0\left(\frac{\mu_n}{R}\rho\right)$ and integrate the results on the interval $[0, R]$, assuming that term-by-term integration is allowed. Using the relations

$$\int_0^l x J_0\left(\frac{\mu_i}{l}x\right) J_0\left(\frac{\mu_j}{l}x\right) dx = \begin{cases} 0, & \text{if } i \neq j, \\ \frac{l^2}{2}[J_0'(\mu_i)]^2, & \text{if } i = j, \end{cases}$$

$$\int_0^x t J_0(t)\, dt = x J_1(x),$$

and

$$\int_0^l x J_0\left(\frac{\mu_n}{l}\,x\right) J_0(kx)\,dx = \frac{\mu_n J_0'(\mu_n) J_0(kl)}{k^2 - \left(\frac{\mu_n}{l}\right)^2}\,, \qquad k \neq \frac{\mu_n}{l},$$

we find that

$$-\frac{P_0 a^2}{\omega T}\left[\frac{a^2 R^2 \mu_n J_0'(\mu_n)}{\omega^2 R^2 - a^2 \mu_n} + \frac{R^2}{\mu_n} J_0'(\mu_n)\right] = ac_n \frac{\mu_n}{R}\frac{R^2}{2}[J_0'(\mu_n)]^2.$$

This yields

$$c_n = \frac{2P_0 a\omega R^3}{T\mu_n^2}\frac{1}{J_0'(\mu_n)(\omega^2 R^2 - a^2\mu_n^2)}\,.$$

Therefore,

$$v(\rho,t) = -\frac{2P_0 a\omega R^3}{T}\sum_{n=1}^{\infty}\frac{\sin\left(a\frac{\mu_n}{R}t\right) J_0\left(\frac{\mu_n}{R}\rho\right)}{\mu_n^2 J_0'(\mu_n)(\omega^2 R^2 - a^2\mu_n^2)}\,.$$

Putting all together, we find the solution of the original problem in the form

$$u(\rho,t) = \frac{P_0}{T}\frac{a^2}{\omega^2}\left[\frac{J_0\left(\frac{\omega\rho}{a}\right)}{J_0\left(\frac{\omega R}{a}\right)} - 1\right]\sin(\omega t) -$$

$$-\frac{2P_0 a\omega R^3}{T}\sum_{n=1}^{\infty}\frac{\sin\left(a\frac{\mu_n}{R}t\right) J_0\left(\frac{\mu_n}{R}\rho\right)}{\mu_n^2 J_0'(\mu_n)(\omega^2 R^2 - a^2\mu_n^2)}\,.$$

Remark 1. The solution of the last problem was obtained under the assumption that the frequency ω of the forcing is not equal to any of the eigenfrequencies $\omega_n = a\mu_n/R$ of the membrane (here μ_n are the positive roots of the equation $J_0(x) = 0$).

Remark 2. The solution of the mixed problem for the wave equation in a disc, formulated mathematically as

$$\begin{cases} u_{tt} = a^2\Delta u, & 0 < \rho < R, \quad t > 0, \\ u(\rho,0) = A\left(1 - \frac{\rho^2}{R^2}\right), & u_t(\rho,0) = 0, & 0 \leq \rho \leq R, \\ |u(0,t)| < \infty, & u(R,t) = 0, & 0 \leq t < \infty, \end{cases} \qquad (2.49)$$

(where A is a constant) can be found in exactly the same way as in the preceding problem. Namely, the solution of problem (2.49) is represented as a series

$$u(\rho,t) = \sum_{n=1}^{\infty}\left[A_n \cos\left(a\frac{\mu_n}{R}t\right) + B_n \sin\left(a\frac{\mu_n}{R}t\right)\right] J_0\left(\frac{\mu_n}{R}\rho\right).$$

From the initial conditions it follows that

$$B_n = 0, \qquad n = 1, 2, \ldots,$$

$$A_n = \frac{2A}{R^2 J_1^2(\mu_n)} \int_0^R \rho(R^2 - \rho^2) J_0\left(\frac{\mu_n}{R}\rho\right) d\rho.$$

It is known that

$$\int_0^x t J_0(t)\, dt = x J_1(x)$$

and

$$\int_0^x t^3 J_0(t)\, dt = 2x^2 J_0(x) + x(x^2 - 4) J_1(x)$$

(both identities are verified by straightforward differentiation and use of the recursion formulas for the Bessel function of the first kind).

Using these identities we finally obtain

$$u(\rho, t) = 8A \sum_{n=1}^{\infty} \frac{J_0\left(\frac{\mu_n}{R}\rho\right)}{\mu_n^3 J_1(\mu_n)} \cos\left(a\frac{\mu_n}{R}t\right),$$

where μ_n are the positive roots of the equation $J_0(x) = 0$.

2.10. The Fourier method. Oscillations of a beam

We have already encountered equations of hyperbolic type of 4th order in the space derivatives when we studied the application of the Fourier and Hankel transformations. As an example of 4th order equation in the one-dimensional case we will consider here the transverse oscillations of a thin beam. The main difference between this situation and the oscillations of a string is that the beam resists bending. Using the laws of mechanics one can show that the oscillations of a beam clamped at one end are described by the equation

$$u_{tt} + a^2 \frac{\partial^4 u}{\partial x^4} = 0 \tag{2.50}$$

where $u(x, t)$ denotes the displacement of the beam (Figure 2.12).

Obviously, the boundary conditions at the clamped end ($x = 0$) are that the rod is stationary and the tangent to it is horizontal ($\frac{\partial u}{\partial x}(0, t) = 0$, whereas at the free end ($x = l$) both the bending moment $M = -E\frac{\partial^2 u}{\partial x^2} J$ (where E is the elasticity modulus of the material of the beam and J is the moment of inertia of the cross section of the beam relative to the horizontal axis) and the tangential force $F = -EJ\frac{\partial^3 u}{\partial x^3}$ must vanish (see Figure 2.12). Notice that in equation (2.50) $a^2 = EJ/\rho s$, where ρ is the density of the material of the beam and s is the area of its cross section).

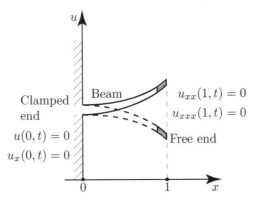

FIGURE 2.12. Oscillations of a beam

To fully determine the motion of a beam clamped at one end we must specify the following initial conditions: the initial displacement (deviation) and the initial velocity in each cross section of the beam, i.e., specify

$$u|_{t=0} = f(x), \quad u_t|_{t=0} = g(x), \qquad 0 \leq x \leq l.$$

Thus, we are led to considering the following mixed problem:

$$\begin{cases} u_{tt} + a^2 u_{xxxx} = 0, \qquad 0 < x < l, \quad t > 0, & (2.51) \\[2mm] u(0,t) = 0, \quad u_x(0,t) = 0, \\ u_{xx}(l,t) = 0, \quad u_{xxx}(l,t) = 0, \qquad t \geq 0, & (2.52) \\[2mm] u(x,0) = f(x), \quad u_t(x,0) = g(x), \qquad 0 \leq x \leq l. & (2.53) \end{cases}$$

We shall solve this problem by the method of separation of variables under the assumption that we are looking for time-periodic oscillations of the beam. Setting $u(x,t) = X(x)T(t)$ and substituting this expression in equation (2.51) we obtain

$$XT'' + a^2 X^{(\mathrm{iv})}T = 0,$$

whence

$$\frac{T''}{a^2 T} = -\frac{X^{(\mathrm{iv})}}{X} = -\lambda.$$

Clearly, in order to guarantee the existence of time-periodic solutions we must take $\lambda > 0$.

Thus, we obtain the following eigenoscillation problem for the function $X(x)$:

$$X^{(\mathrm{iv})} - \lambda X = 0, \tag{2.54}$$

with the boundary conditions

$$X(0) = 0, \quad X'(0) = 0, \quad X''(l) = 0, \quad X'''(l) = 0. \tag{2.55}$$

The general solution of equation (2.54) has the form

$$X(x) = A\cosh(\sqrt[4]{\lambda}\,x) + B\sinh(\sqrt[4]{\lambda}\,x) + C\cos(\sqrt[4]{\lambda}\,x) + D\sin(\sqrt[4]{\lambda}\,x).$$

From the conditions $X(0) = 0$, $X'(0) = 0$ we find that $A + C = 0$ and $B + D = 0$, and so

$$X(x) = A[\cosh(\sqrt[4]{\lambda}\,x) - \cos(\sqrt[4]{\lambda}\,x)] + B[\sinh(\sqrt[4]{\lambda}\,x) - \sin(\sqrt[4]{\lambda}\,x)].$$

The boundary conditions (2.55) at the right end of the beam yield

$$\begin{cases} A[\cosh(\sqrt[4]{\lambda}\,l) + \cos(\sqrt[4]{\lambda}\,l)] + B[\sinh(\sqrt[4]{\lambda}\,l) + \sin(\sqrt[4]{\lambda}\,l)] = 0, \\ A[\sinh(\sqrt[4]{\lambda}\,l) - \sin(\sqrt[4]{\lambda}\,l)] + B[\cosh(\sqrt[4]{\lambda}\,l) + \cos(\sqrt[4]{\lambda}\,l)] = 0. \end{cases} \tag{2.56}$$

The homogeneous system (2.56) for the unknowns A and B has a nontrivial solution if and only if its determinant is equal to zero:

$$\begin{vmatrix} \cosh(\sqrt[4]{\lambda}l) + \cos(\sqrt[4]{\lambda}l) & \sinh(\sqrt[4]{\lambda}l) + \sin(\sqrt[4]{\lambda}l) \\ \sinh(\sqrt[4]{\lambda}x) - \sin(\sqrt[4]{\lambda}x) & \cosh(\sqrt[4]{\lambda}x) + \cos(\sqrt[4]{\lambda}x) \end{vmatrix} = 0. \tag{2.57}$$

From (2.57) we obtain an algebraic equation for the computation of the eigenvalues of our problem:

$$\sinh^2(\sqrt[4]{\lambda}l) - \sin^2(\sqrt[4]{\lambda}l) = \cosh^2(\sqrt[4]{\lambda}x) +$$

$$+ 2\cosh(\sqrt[4]{\lambda}x)\cos(\sqrt[4]{\lambda}x) + \cos^2(\sqrt[4]{\lambda}x). \tag{2.58}$$

Denoting $\mu = \sqrt[4]{\lambda}\,l$ and using the identity $\sinh^2 x + 1 = \cosh^2 x$, we obtain from (2.58) the equation

$$\cosh\mu \cdot \sinh\mu = -1. \tag{2.59}$$

This equation can be solved graphically (Figure 2.13).

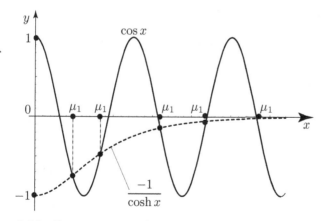

FIGURE 2.13. Determination of the roots of the equation $\cosh\mu \cdot \sinh\mu = -1$

One obtains the values

$$\mu_1 = 1.857, \qquad \mu_2 = 4.694, \qquad \mu_3 = 7.854,$$

$$\mu_n \approx \frac{\pi}{n}(2n - 1) \qquad \text{for } n > 3.$$

Further, the function $T(t)$ satisfies the equation

$$T'' + \lambda_n a^2 T = 0.$$

Its general solution has the form

$$T_n(t) = A_n \cos(a\sqrt{\lambda_n} t) + B_n \sin(a\sqrt{\lambda_n} t) =$$

$$= A_n \cos\left(a \frac{\mu_n^2}{l^2} t\right) + B_n \sin\left(a \frac{\mu_n^2}{l^2} t\right),$$

where A_n and B_n are arbitrary constants.

Therefore, the solution "atoms" of the problem (2.51), (2.52) are provided by the functions

$$u_n(x,t) = \left[A_n \cos\left(a \frac{\mu_n^2}{l^2} t\right) + B_n \sin\left(a \frac{\mu_n^2}{l^2} t\right)\right] X_n(x),$$

where

$$X_n(x) = \frac{(\sinh\mu_n + \sin\mu_n)\left[\cosh\left(\frac{\mu_n}{l} x\right) - \cos\left(\frac{\mu_n}{l} x\right)\right]}{\sinh\mu_n + \sin\mu_n} -$$

$$- \frac{(\cosh\mu_n + \cos\mu_n)\left[\sinh\left(\frac{\mu_n}{l} x\right) - \sin\left(\frac{\mu_n}{l} x\right)\right]}{\sinh\mu_n + \sin\mu_n}.$$

According to the general theory of Sturm-Liouville problems, the eigenfunctions $\{X_n(x)\}_{n=1}^{\infty}$ form a complete orthogonal system of functions on the segment $[0, l]$. Then the solution of the problem (2.51)–(2.53) is given by the series

$$u(x,t) = \sum_{n=1}^{\infty}\left[A_n \cos\left(a \frac{\mu_n^2}{l^2} t\right) + B_n \sin\left(a \frac{\mu_n^2}{l^2} t\right)\right] X_n(x),$$

where the coefficients A_n and B_n are determined from the initial conditions via the formulas

$$A_n = \frac{\int_0^l f(x)X_n(x)\,dx}{\|X_n(x)\|^2} \qquad B_n = \frac{\int_0^l g(x)X_n(x)\,dx}{a \frac{\mu_n^2}{l}\|X_n(x)\|^2},$$

where $\|X_n(x)\|^2 = \int_0^l X_n^2(x)\,dx$.

2.11. The perturbation method

In this section we will consider briefly the perturbation method, which lies at the basis of the solution of many problems of mathematical physics (see, e.g., [11], [13], [23]). This method enables one to reduce the solution of some complex problems to a series of simple problems according to the scheme shown in Figure 2.14.

Suppose we want to solve the Cauchy problem for the equation with variable coefficients

$$\begin{cases} u_{tt} = a^2 u_{xx} + c(x,t)u, & -\infty < x < \infty, \quad t > 0, & (2.60) \\ u(x,0) = f(x), \quad u_t(x,0) = g(x), & -\infty < x < \infty. & (2.61) \end{cases}$$

Problem (2.60), (2.61) can be solved starting from the Cauchy problem for a simpler equation.

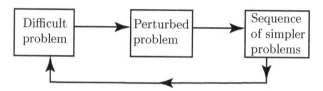

FIGURE 2.14. Block diagram of the perturbation method

To do this, we will consider the perturbed problem

$$\begin{cases} u_{tt} = a^2 u_{xx} + \varepsilon c(x,t)u, & -\infty < x < \infty, \quad t > 0, & (2.62) \\ u(x,0) = f(x), \quad u_t(x,0) = g(x), & -\infty < x < \infty. & (2.63) \end{cases}$$

and seek its solution as a series

$$u(x,t) = u_0(x,t) + \varepsilon u_1(x,t) + \varepsilon^2 u_2(x,t) + \dots,$$

where $0 \le \varepsilon \le 1$ and the functions $\{u_k(x,t)\}_{k=0}^{\infty}$ need to be determined.

Substituting the proposed form of the solution in equation (2.60) we obtain

$$\frac{\partial^2}{\partial t^2}[u_0(x,t) + \varepsilon u_1(x,t) + \varepsilon^2 u_2(x,t) + \dots] =$$

$$= a^2 \frac{\partial^2}{\partial x^2}[u_0(x,t) + \varepsilon u_1(x,t) + +\varepsilon^2 u_2(x,t) + \dots] +$$

$$+ \varepsilon c(x,t)[u_0(x,t) + \varepsilon u_1(x,t) + \varepsilon^2 u_2(x,t) + \dots]. \quad (2.64)$$

Performing a number of simple algebraic transformation and equating the coefficients of like powers of ε, we obtain the following sequence of equations:

$$\varepsilon^0: \quad \frac{\partial^2 u_0}{\partial t^2} = a^2 \frac{\partial^2 u_0}{\partial x^2}, \quad (2.65)$$

$$\varepsilon^1 : \qquad \frac{\partial^2 u_1}{\partial t^2} = a^2 \frac{\partial^2 u_1}{\partial x^2} + c(x,t) u_0(x,t), \qquad (2.66)$$

$$\varepsilon^2 : \qquad \frac{\partial^2 u_2}{\partial t^2} = a^2 \frac{\partial^2 u_2}{\partial x^2} + c(x,t) u_1(x,t), \qquad (2.67)$$

. . . .

Similarly, for the initial conditions of problem (2.62), (2.63) we obtain series in powers of ε:

$$f(x) = u_0(x,0) + \varepsilon u_1(x,0) + \varepsilon^2 u_2(x,0) + \dots , \qquad (2.68)$$

$$g(x) = \frac{\partial u_0}{\partial t}(x,0) + \varepsilon \frac{\partial u_1}{\partial t}(x,0) + \varepsilon^2 \frac{\partial u_2}{\partial t}(x,0) + \dots \qquad (2.69)$$

These relations give the initial conditions for the problems (2.65), (2.66), and so on. Namely,

$$\begin{cases} u_0(x,0) = f(x), & \dfrac{\partial u_0}{\partial t} = 0, \\[2mm] u_n(x,0) = 0, & \dfrac{\partial u_n}{\partial t} = 0, \qquad n \geq 1. \end{cases}$$

Therefore, we have the sequence of Cauchy problems

$$\begin{cases} \dfrac{\partial^2 u_0}{\partial t^2} = a^2 \dfrac{\partial^2 u_0}{\partial x^2}, & -\infty < x < \infty, \quad t > 0, \\[2mm] u_0(x,0) = f(x), & \dfrac{\partial u_0}{\partial t}(x,0) = g(x), \qquad -\infty < x < \infty. \end{cases} \qquad (2.70)$$

$$\begin{cases} \dfrac{\partial^2 u_1}{\partial t^2} = a^2 \dfrac{\partial^2 u_1}{\partial x^2} + c(x,t) u_0(x,t), & -\infty < x < \infty, \quad t > 0, \\[2mm] u_1(x,0) = 0, & \dfrac{\partial u_1}{\partial t}(x,0) = 0, \qquad -\infty < x < \infty. \end{cases} \qquad (2.71)$$

. . . .

The solution $u_0(x,t)$ of problem (2.70) is found by means of d'Alembert's formula, so that it is a known function. The solution of problem (2.71) is given by the formula

$$u_1(x,t) = \frac{1}{2a} \int_0^t d\tau \int_{x-a(t-\tau)}^{x+a(t-\tau)} c(\zeta,\tau) u_0(\zeta,\tau) \, d\zeta.$$

Thus, we found the solutions of problems (2.70) and (2.71).

The function $u_1(x,t)$ is often referred to as the *first-order perturbation* of the function $u_0(x,t)$. Adding $u_0(x,t)$ and $u_1(x,t)$ we obtain an approximate solution of problem (2.60), (2.61).

To find the next-order perturbation $u_2(x,t)$ we must solve the Cauchy problem for equation (2.67):

$$\begin{cases} \dfrac{\partial^2 u_2}{\partial t^2} = a^2 \dfrac{\partial^2 u_2}{\partial x^2} + c(x,t)u_1(x,t), & -\infty < x < \infty, \quad t > 0, \\[2mm] u_2(x,0) = 0, \qquad \dfrac{\partial u_2}{\partial t}(x,0) = 0, & -\infty < x < \infty. \end{cases}$$

In a similar manner one can find the other functions u_n, $n = 3, 4, \dots$. The exact solution of problem (2.60), (2.61) is given by the series $u(x,t) = \sum_{n=0}^{\infty} u_n(x,t)$.

Now let us give an example of a *nonlinear hyperbolic problem*. Suppose we are required to solve the Cauchy problem

$$u_{tt} = a^2 u_{xx} + u u_x, \qquad -\infty < x < \infty, \quad t > 0, \tag{2.72}$$

$$\begin{cases} u(x,0) = f(x) + g(x), \\ u_t(x,0) = a[g'(x) - f'(x)], \end{cases} \quad \infty < x < \infty. \tag{2.73}$$

We will seek the solution of this problem starting from the following perturbed problem [13, p. 66, no. 2.20] (and then take $\varepsilon = 1$)

$$u_{tt} = a^2 u_{xx} + \varepsilon u u_x, \qquad -\infty < x < \infty, \quad t > 0, \tag{2.74}$$

$$\begin{cases} u(x,0) = f(x) + g(x), \\ u_t(x,0) = a[g'(x) - f'(x)], \end{cases} \quad \infty < x < \infty. \tag{2.75}$$

Setting $u(x,t) = \sum_{n=0}^{\infty} \varepsilon^n u_n(x,t)$, let us substitute this series in equation (2.74). We obtain

$$\sum_{n=0}^{\infty} \varepsilon^n \frac{\partial^2 u_n}{\partial t^2} = a^2 \sum_{n=0}^{\infty} \varepsilon^n \frac{\partial^2 u_n}{\partial x^2} + \varepsilon \left(\sum_{n=0}^{\infty} \varepsilon^n u_n \right) \left(\sum_{n=0}^{\infty} \varepsilon^n \frac{\partial u_n}{\partial x} \right),$$

or, in expanded form,

$$\frac{\partial^2 u_0}{\partial t^2} - a^2 \frac{\partial^2 u_0}{\partial x^2} + \varepsilon \left(\frac{\partial^2 u_1}{\partial t^2} - a^2 \frac{\partial^2 u_1}{\partial x^2} \right) + \dots =$$

$$= \varepsilon (u_0 + \varepsilon u_1 + \dots) \left(\frac{\partial u_0}{\partial x} + \varepsilon \frac{\partial u_1}{\partial x} + \dots \right).$$

If we now equate the coefficients of like powers of ε we obtain a sequence of Cauchy problems:

$$\varepsilon^0 : \begin{cases} \dfrac{\partial^2 u_0}{\partial t^2} = a^2 \dfrac{\partial^2 u_0}{\partial x^2}, & -\infty < x < \infty, \quad t > 0, \\[2mm] u_0(x,0) = f(x) + g(x), & \\[2mm] \dfrac{\partial u_0}{\partial t}(x,0) = a[g'(x) - f'(x)], & -\infty < x < \infty, \end{cases} \tag{2.76}$$

$$\varepsilon^1 : \quad \begin{cases} \dfrac{\partial^2 u_1}{\partial t^2} = a^2 \dfrac{\partial^2 u_1}{\partial x^2} + u_0(x,t)\dfrac{\partial u_0}{\partial x}(x,t), \quad -\infty < x < \infty, \quad t > 0, \\ u_1(x,0) = f(x) + g(x), \\ \qquad\qquad\qquad\qquad\qquad\qquad -\infty < x < \infty, \\ \dfrac{\partial u_1}{\partial t}(x,0) = 0, \end{cases}$$

$$(2.77)$$

. . . .

The solution of problem (2.76) is found by means of d'Alembert's formula. After simple manipulations we get

$$u_0(x,t) = g(x + at) + f(x - at).$$

Now let us find $u_1(x,t)$, solving the linear problem

$$\begin{cases} \dfrac{\partial^2 u_1}{\partial t^2} = a^2 \dfrac{\partial^2 u_1}{\partial x^2} + [g(x + at) + f(x - at)][g'(x + at) + f'(x - at)], \\ \qquad\qquad -\infty < x < \infty, \quad t > 0, \\ u_1(x,0) = 0, \qquad -\infty < x < \infty, \\ \dfrac{\partial u_1}{\partial t}(x,0) = 0, \qquad -\infty < x < \infty. \end{cases}$$

$$(2.78)$$

The solution of problem (2.78) is written in the form

$$u_1(x,t) = \frac{1}{2a}\int_0^t d\tau \int_{x-a(t-\tau)}^{x+a(t-\tau)} [g(\zeta + a\tau) + f(\zeta - a\tau)][g'(\zeta + a\tau) + f'(\zeta - a\tau)]d\zeta =$$

$$\frac{1}{2a}\int_0^t d\tau \int_{x-a(t-\tau)}^{x+a(t-\tau)} \frac{1}{2}\frac{\partial}{\partial\zeta}[g(\zeta+a\tau)]^2 d\zeta + \frac{1}{2a}\int_0^t d\tau \int_{x-a(t-\tau)}^{x+a(t-\tau)} \frac{1}{2}\frac{\partial}{\partial\zeta}[f(\zeta-a\tau)]^2 d\zeta +$$

$$+ \frac{1}{2a}\int_0^t \frac{\partial}{\partial\zeta}[g(\zeta + a\tau)f(\zeta - a\tau)]d\zeta =$$

$$= \frac{1}{4a}\int_0^t [g^2(x + at) - g^2(x - at + 2a\tau)]d\tau +$$

$$+ \frac{1}{4a}\int_0^t [f^2(x + at - 2a\tau) - f^2(x - at)]d\tau +$$

$$+ \frac{1}{4a}\int_0^t [2g(x + at)f(x + at - 2a\tau) - 2g(x - at + 2a\tau)f(x - at)]d\tau.$$

Therefore, we found the function $u_1(x,t)$.

An approximate solution of the nonlinear problem (2.720, (2.73) is provided by $u(x,t) = u_0(x,t) + u_1(x,t)$, and the exact solution is given by the series

$$u(x,t) = \sum_{n=0}^{\infty} u_n(x,t).$$

2.12. Problems for independent study

1. Find the transversal oscillations of an infinite string if the initial displacement and the initial velocity are equal to zero and xe^{-x^2}, respectively. Construct the graph of the solution at different moments of time.

2. Solve the following problem for a semi-infinite string:
$$\begin{cases} u_{tt} = u_{xx}, & 0 < x < \infty, \quad t > 0, \\ u(0,t) = 0, & t \geq 0, \\ u(x,0) = -xe^{-x^2}, & u_t(x,0) = 0, \quad 0 \leq x < \infty. \end{cases}$$

3. Solve the Cauchy problem for the infinite string
$$\begin{cases} u_{tt} = u_{xx} + e^{-t}, & -\infty < x < \infty, \quad t > 0, \\ u(x,0) = \sin x, \quad u_t(x,0) = \cos x, & -\infty < x < \infty. \end{cases}$$

4. Solve the Cauchy problem for the infinite membrane
$$\begin{cases} u_{tt} = u_{xx} + u_{yy} + t\sin y, & -\infty < x, y < \infty, \quad t > 0, \\ u(x,y,0) = x^2, \quad u_t(x,y,0) = \sin y, & -\infty < x, y < \infty. \end{cases}$$

5. Solve the Cauchy problem in three-dimensional space
$$\begin{cases} u_{tt} = 8(u_{xx} + u_{yy} + u_{zz}) + t^2 x^2, & -\infty < x, y, z < \infty, \quad t > 0, \\ u(x,y,z,0) = y^2, \quad u_t(x,y,z,0) = z^2, & -\infty < x, y, z < \infty. \end{cases}$$

6. Solve the problem for the semi-infinite string
$$\begin{cases} u_{tt} = a^2 u_{xx}, & 0 < x < \infty, \quad t > 0, \\ u(0,t) = 0, & t \geq 0, \\ u(x,0) = f(x), \quad u_t(x,0) = g(x), & 0 \leq x < \infty. \end{cases}$$

7. Solve the Cauchy problem for the infinite membrane
$$\begin{cases} u_{tt} = a^2(u_{xx} + u_{yy}) + f(x,y,t), & -\infty < x, y < \infty, \quad t > 0, \\ u(x,y,0) = 0, \quad u_t(x,y,0) = 0, & -\infty < x, y < \infty. \end{cases}$$

8. Solve the Cauchy problem in three-dimensional space
$$\begin{cases} u_{tt} = a^2(u_{xx} + u_{yy} + u_{zz}) + f(x,y,z,t), & -\infty < x, y, z < \infty, \quad t > 0, \\ u(x,y,z,0) = 0, \quad u_t(x,y,z,0) = 0, & -\infty < x, y, z < \infty. \end{cases}$$

9. Find the radially symmetric transversal oscillations of an unbounded plate by solving the Cauchy problem

$$\begin{cases} u_{tt} + b^2 \Delta^2 u = 0, & 0 < \rho < \infty, \quad t > 0, \\ u|_{t=0} = Ae^{-\rho^2/a^2}, & u_t|_{t=0} = 0, \quad 0 \leq \rho < \infty. \end{cases}$$

10. At the point $x = 0$ along the axis of a unbounded rod ($x > 0$) a force $f(t) = \cos t$ is applied starting at time $t = 0$, at which the rod is at rest. Find the longitudinal oscillations of the rod.

11. The ends $x = 0$ and $x = l$ of a string are rigidly clamped. The initial velocity is given by $u_t(x, 0) = A \sin \left(\frac{2\pi x}{l} \right)$ for $0 \leq x \leq l$. Find the displacement at time $t > 0$ if one knows that the initial displacement is $u(x, 0) = 0$.

12. An insulated electrical wire $0 \leq x \leq l$ is charged to a potential $v_0 = $ const. At the initial moment of time the extremity $x = 0$ is grounded, while the extremity $x = l$ remains isolated. Find the voltage distribution in the wire if the selfinductance and capacity per unit of length of the wire are known. Resistance is neglected.

13. Find the longitudinal oscillations of a rod $0 \leq x \leq l$ whose left extremity is rigidly clamped, while at the right extremity a force $F(t) = At$, $A = $ const, is applied starting at time $t = 0$.

14. Find the oscillations of a string $0 \leq x \leq l$ with rigidly clamped ends, and subject to the force $F(x, t) = xt$, applied starting at time $t = 0$.

15. Consider a wire of length l traversed by a variable current, and assume that there is no current leakage and ohmic resistance can be neglected. The initial current through the wire is given by $E = E_0 \sin \left(\frac{\pi x}{2l} \right)$, the left extremity of the wire ($x = 0$) is insulated and the right extremity ($x = l$) is grounded. Find the intensity of the current.

16. Solve the mixed problem on the free oscillations of a rectangular membrane ($0 \leq x \leq \pi, 0 \leq y \leq \pi$), with the initial displacement $u(x, y, 0) = Axy(x - \pi)(y - \pi)$ and the initial velocity $u_t(x, y, 0) = 0$.

17. Solve the following mixed problem on forced oscillations of a membrane:

$$\begin{cases} u_{tt} = u_{xx} + u_{yy} + \sin t \sin x \sin y, & 0 < x, y < \pi, \quad t > 0, \\ u|_{x=0} = u|_{x=\pi} = u|_{y=0} = u|_{y=\pi} = 0, & t \geq 0, \\ u(x, y, 0) = 0, \quad u_t(x, y, 0) = 0, & 0 \leq x, y \leq \pi. \end{cases}$$

18. Solve the following mixed problem on the oscillations of a membrane:

$$\begin{cases} u_{tt} = c^2 (u_{xx} + u_{yy}), & 0 < x < a, \quad 0 < y < b, \quad t > 0, \\ u(0, y, t) = \sin \left(\frac{\pi y}{b} \right), & u(a, y, t) = 0, \quad 0 \leq y \leq b, \quad t \geq 0, \\ u(x, 0, t) = 0, \quad u(x, b, t) = 0, & 0 \leq x \leq a, \quad t \geq 0, \\ u(x, y, 0) = 0, \quad u_t(x, y, 0) = 0, & 0 \leq x \leq a, \quad 0 \leq y \leq b. \end{cases}$$

19. A homogeneous circular membrane of radius b, clamped along it boundary, is at rest under tension T_0. A normal pressure of magnitude P per unit of area is applied to the membrane at $t = 0$. Find the radial oscillations of the membrane.

20. Solve the following mixed problem for the wave equation in an infinite cylinder $(-\infty < z < \infty)$:

$$\begin{cases} u_{tt} = \Delta u + J_0 \left(\dfrac{\mu_1}{a} \rho \right), & 0 < \rho < a, \quad t > 0, \\ u|_{t=0} = 0, \quad u_t|_{t=0} = 0, & 0 \le \rho \le a, \\ u|_{\rho=a} = 0, & t \ge 0, \end{cases}$$

where μ_1 is the first positive root of the Bessel function $J_0(x)$.

21. Solve the Cauchy problem for the wave equation in a bounded cylinder:

$$\begin{cases} u_{tt} = \Delta u, & 0 < \rho < a, \quad 0 < z < b, \quad t > 0, \\ u|_{t=0} = J_0 \left(\dfrac{\mu_2}{a} \rho \right) \cos \left(\dfrac{3\pi z}{b} \right), \quad u_t|_{t=0} = 0, & 0 \le \rho \le a, \quad 0 \le z \le b \\ \left. \dfrac{\partial u}{\partial z} \right|_{z=0} = \left. \dfrac{\partial u}{\partial z} \right|_{z=b} = 0, \quad \left. \dfrac{\partial u}{\partial \rho} \right|_{\rho=a} = 0 & t \ge 0, \end{cases}$$

where μ_2 is the second positive root of the Bessel function $J_1(x)$.

22. Solve the following mixed problem on the forced oscillations of a membrane:

$$\begin{cases} u_{tt} = u_{xx} + u_{yy} + t \cos(3x) \cos(4y), & 0 < x, y < \pi, \quad t > 0, \\ u_x|_{x=0} = u_x|_{x=\pi} = u_y|_{y=0} = u_y|_{y=\pi} = 0 & t \ge 0 \\ u|_{t=0} = 0, \quad u_t|_{t=0} = 0, & 0 \le x, y \le \pi. \end{cases}$$

23. Solve the following mixed problem on the oscillations of a beam clamped at its extremities:

$$\begin{cases} u_{tt} + u_{xxxx} = 0, & 0 < x < l, \quad 0 < t < \infty, \\ u(0,t) = 0, \quad u(l,t) = 0, \quad u_{xx}(0,t) = 0, \quad u_{xx}(0,l) = 0, & t \ge 0, \\ u(x,0) = 0, \quad u_t(x,0) = \sin \left(\dfrac{2\pi}{l} x \right), & 0 \le x \le l. \end{cases}$$

24. Determine up to terms of order ε^2 the solution of the Cauchy problem

$$\begin{cases} u_{tt} - u_{xx} + u = \varepsilon u^3, & -\infty < x < \infty, \quad t > 0, \\ u(x,0) = \cos x, \quad u_t(x,0) = 0, & -\infty < x < \infty. \end{cases}$$

25. Establish the uniqueness of the solution of the following mixed problem on the oscillations of a membrane:

$$\begin{cases} u_{tt} = a^2(u_{xx} + u_{yy}) + f(x,y,t), & 0 < x < a,\ 0 < y < b,\ 0 < t < T, \\ u|_{x=0} = f_1(y,t), \quad u|_{x=a} = f_2(y,t), & 0 \le y \le b,\ 0 \le t \le T, \\ u|_{y=0} = g_1(x,t), \quad u|_{y=b} = g_2(x,t), & 0 \le x \le a,\ 0 \le t \le T, \\ u|_{t=0} = F(x,y), \quad u_t|_{t=0} = G(x,y), & 0 \le x \le a,\ 0 \le y \le b, \end{cases}$$

under the assumption that the function $u(x,y,z)$ is twice continuously differentiable in the closed domain $0 \le x \le a,\ 0 \le y \le b,\ 0 \le t \le T$.

26. Solve the following mixed problem on the oscillations of a circular membrane:

$$\begin{cases} u_{tt} = c^2 \Delta u, & 0 < \rho < a,\quad 0 \le \varphi < 2\pi,\quad t > 0, \\ u(a,\varphi,t) = 0, & 0 \le \varphi 2 \le \pi,\quad t \ge 0 \\ u|_{t=0} = 0,\quad u_t|_{t=0} = J_3\left(\dfrac{\mu_5}{a}\rho\right)\cos(3\varphi), & 0 \le \rho \le a,\quad 0 \le \varphi \le 2\pi, \end{cases}$$

where μ_5 is the fifth positive root of the Bessel function $J_3(x)$ and

$$\Delta u = \frac{1}{\rho}\frac{\partial}{\partial \rho}\left(\rho \frac{\partial u}{\partial \rho}\right) + \frac{1}{\rho^2}\frac{\partial^2 u}{\partial \varphi^2}.$$

27. Solve the following mixed problem on the oscillations of a membrane (with nonzero boundary condition):

$$\begin{cases} u_{tt} = c^2 \Delta u, & 0 < \rho < a,\quad 0 \le \varphi < 2\pi, t > 0, \\ u(a,\varphi,t) = A\cos\varphi, & 0 \le \varphi \le \pi,\quad t \ge 0, \\ u(\rho,\varphi,0) = 0,\quad u_t(\rho,\varphi,0) = 0, & 0 \le \rho \le a,\quad 0 \le \varphi \le 2\pi, \end{cases}$$

28. Find the oscillations of a circular membrane $(0 \le \rho \le a)$, with null boundary conditions, generated by the motion of its boundary according to the law

$$u(a,\varphi,t) = A\cos(\omega t), \qquad A = \text{const.}$$

29. Solve the following mixed problem on the oscillations of a membrane (with nonzero boundary conditions):

$$\begin{cases} u_{tt} = c^2 \Delta u, & 0 < \rho < a,\quad 0 \le \varphi < 2\pi,\quad t > 0, \\ u(a,\varphi,t) = A\sin(2\varphi), & 0 \le \varphi \le \pi,\quad t \ge 0, \\ u(\rho,\varphi,0) = 0,\quad u_t(\rho,\varphi,0) = 0, & 0 \le \rho \le a,\quad 0 \le \varphi \le 2\pi. \end{cases}$$

30. Find the oscillations of a circular membrane $(0 \le \rho \le a)$, with null boundary conditions, generated by the motion of its boundary according to the law

$$u(a,\varphi,t) = A\sin(\omega t), \qquad A = \text{const.}$$

31. Solve the following mixed problem for the wave equation in a bounded cylinder:

$$\begin{cases} u_{tt} = c^2 \Delta u + t J_0 \left(\dfrac{\mu_1}{a} \rho \right) \sin \left(\dfrac{2\pi}{h} z \right), & 0 < \rho < a, \quad 0 < z < h, t > 0, \\ u|_{z=0} = 0, \quad u|_{z=h} = 0, \quad u|_{\rho=a} = 0, & t \geq 0, \\ u|_{t=0} = 0, \quad u_t|_{t=0} = 0, \end{cases}$$

where μ_1 is the first positive root of the equation $J_0(x) = 0$.

32. Find the oscillations of a gas in a spherical container, generated by oscillations of its walls that start at $t = 0$. It is assumed that the velocity of the particles are directed along the radii of the container and have magnitude equal to $\frac{1}{2} t^2 P_n(\cos \theta)$, where $P_n(x)$ is the Legendre polynomial of degree n.

33. Show that if the functions $f(x)$, $u_0(x)$ and $u_1(x)$ are harmonic in the whole Euclidean space $\mathbf{R}^n$ and $g(t)$ is a continuously differentiable function for $t \geq 0$, then the solution of the Cauchy problem

$$\begin{cases} u_{tt} = c^2 \Delta u + f(x)g(t), & x \in \mathbf{R}^n, \quad t > 0, \\ u|_{t=0} = u_0(x), \quad u_t|_{t=0} = u_1(x), & x \in \mathbf{R}^n, \end{cases}$$

is given by the formula

$$u(x, t) = u_0(x) + t u_1(x) + f(x) \int_0^t (t - \tau) g(\tau) \, d\tau.$$

34. Find the oscillations of an infinite circular cylinder of radius a, generated by the motion of its lateral surface according to the law $A \sin(\omega t) \cos(n\varphi)$, where $A = \text{const}$ and n is an arbitrary fixed positive integer.

35. Prove that if $u_f(x, t)$ is the solution of the Cauchy problem

$$\begin{cases} u_{tt} = a^2 \Delta u, & x \in \mathbf{R}^n, \quad t > 0, \\ u|_{t=0} = 0, \quad u_t|_{t=0} = f(x), & x \in \mathbf{R}^n, \end{cases}$$

then the solution of the Cauchy problem

$$\begin{cases} u_{tt} = a^2 \Delta u, & x \in \mathbf{R}^n, \quad t > 0, \\ u|_{t=0} = f(x), \quad u_t|_{t=0} = 0, & x \in \mathbf{R}^n, \end{cases}$$

is given by the formula $u(x, t) = \partial u_f(x, t)/\partial t$.

36. Solve the nonlinear Cauchy problem

$$\begin{cases} u_{tt} = a^2 \Delta u + (\nabla u)^2 - u_t^2, & x \in \mathbf{R}^3, \quad t > 0, \\ u(x, 0) = 0, \quad u_t(x, 0) = f(x), & x \in \mathbf{R}^3. \end{cases}$$

37. Show that the function $u(x,t)$ is the solution of the Cauchy problem

$$\begin{cases} u_{tt} + \Delta^2 u = 0, & x \in \mathbf{R}^n, \quad t > 0, \\ u|_{t=0} = f(x), & u_t|_{t=0} = 0, \quad x \in \mathbf{R}^n. \end{cases}$$

if and only if the function

$$v(x,t) = u(x,t) + i \int_0^t \Delta u(x,\tau)\, d\tau$$

is the solution of the Cauchy problem for the homogeneous Schrödinger equation

$$\begin{cases} v_t = i\Delta v, & x \in \mathbf{R}^n, \quad t > 0, \\ v|_{t=0} = f(x), & x \in \mathbf{R}^n. \end{cases}$$

38. Let the function $u(x,t)$ be the solution of the Cauchy problem for the homogeneous Schrödinger equation

$$\begin{cases} u_t = i\Delta u, & x \in \mathbf{R}^n, \quad t > 0, \\ u|_{t=0} = f(x), & x \in \mathbf{R}^n, \end{cases}$$

where $f(x)$ is a real-valued function. Find the solution of the Cauchy problem

$$\begin{cases} v_{tt} + \Delta v = 0, & x \in \mathbf{R}^n, \quad t > 0, \\ v|_{t=0} = f(x), & v_t|_{t=0} = 0, \quad x \in \mathbf{R}^n. \end{cases}$$

39. Let the function $f(x,t)$ be biharmonic in the variable x for each fixed $t \geq 0$ (i.e., $\Delta^2 f = 0$). Find the solution of the Cauchy problem

$$\begin{cases} u_{tt} + \Delta^2 u = f(x,t), & x \in \mathbf{R}^n, \quad t > 0, \\ u|_{t=0} = 0, & u_t|_{t=0} = 0, \quad x \in \mathbf{R}^n. \end{cases}$$

40. Find the solution of the Cauchy problem

$$\begin{cases} u_{tt} + \Delta^2 u = 0, & x \in \mathbf{R}^n, \quad t > 0, \\ u|_{t=0} = u_0(x), & u_t|_{t=0} = u_1(x), \quad x \in \mathbf{R}^n, \end{cases}$$

where $u_0(x)$ and $u_1(x)$ are biharmonic functions (i.e., $\Delta^2 u_0 = 0$, $\Delta^2 u_1 = 0$).

2.13. Answers

1. $u = \dfrac{1}{2a} e^{-(x^2+a^2t^2)} \sinh(2xat)$.

2. $u = e^{-(x^2+t^2)}[x \cosh(2xt) - t \sinh(2xt)]$.

3. $u = t - 1 + e^{-t} + \sin(x + t)$.

4. $u = x^2 + t^2 + t \sin y$.

5. $u = y^2 + tz^2 + 8t^2 + \dfrac{8}{3} t^3 + \dfrac{1}{12} t^4 x^2 + \dfrac{2}{45} t^6$.

6. $u = \dfrac{f(x + at) + f(x - at)}{2} + \dfrac{1}{2a} \displaystyle\int_{|x-at|}^{x+at} g(\xi)\, d\xi$.

7. $u(x, y, z) = \dfrac{1}{2\pi a} \displaystyle\iiint_{\{r \le a(t-\tau)\}} \dfrac{f(\alpha, \beta, \gamma)}{\sqrt{a^2(t - \tau)^2 - r^2}}\, d\alpha\, d\beta$,

$$r = \sqrt{(x - \alpha)^2 + (y - \beta)^2}\,.$$

8. $u(x, y, z, t) = \dfrac{1}{4\pi a^2} \displaystyle\iiint_{\{r \le at\}} \dfrac{f(\alpha, \beta, \gamma, t - r/a)}{r}\, d\alpha\, d\beta\, d\gamma$,

$$r = \sqrt{(x - \alpha)^2 + (y - \beta)^2 + (z - \gamma)^2}\,.$$

9. $u(r, t) = \dfrac{Ae^{-R^2/(1+\tau^2)}}{1 + \tau^2} \left[\cos\left(\dfrac{R^2\tau}{1 + \tau^2} \right) + \tau \sin\left(\dfrac{R^2\tau}{1 + \tau^2} \right) \right]$,

where $\tau = 4bt/a^2$ and $R = r/a$.

10. $u = \begin{cases} 0, & \text{if } t < x/a, \\ ka \sin(t - k/a), & \text{if } t > x/a, \end{cases}$

where k is the Young modulus of the material of the rod.

11. $u = \dfrac{lA}{2\pi a} \sin\left(\dfrac{2\pi a}{l} t \right) \sin\left(\dfrac{2\pi x}{l} \right)$.

12. $u = \dfrac{4v_0}{\pi} \displaystyle\sum_{n=1}^{\infty} \dfrac{1}{2n + 1} \sin\left[\dfrac{(2n + 1)\pi x}{2l} \right] \cos\left[\dfrac{(2n + 1)\pi t}{2l\sqrt{CL}} \right]$.

13. $u = \dfrac{A}{ES} xt + \displaystyle\sum_{n=1}^{\infty} a_n \sin\left[\dfrac{(2n + 1)\pi x}{2l} \right] \cos\left[\dfrac{(2n + 1)\pi at}{2l} \right]$,

where

$$a_n = -\dfrac{4}{\pi a(2n + 1)} \int_0^l \dfrac{Az}{ES} \sin\left[\dfrac{(2n + 1)\pi z}{2l} \right] dz,$$

and E and S are the elasticity modulus and the area of the cross section of the rod, respectively.

14. $u = \sum_{n=1}^{\infty} u_n(t)\sin\left(\frac{n\pi}{l}x\right)$, where $u_n(t) = \frac{2l^2(-1)^{n+1}}{\rho a n^2 \pi^2}\int_0^t \tau\sin\left[\frac{n\pi a}{l}(t-\tau)\right]d\tau$.

15. $u = -E_0\sqrt{\frac{C}{L}}\cos\left(\frac{\pi}{2l}x\right)\sin\left(\frac{\pi}{2l\sqrt{LC}}t\right)$, where L and C are the inductance

and the capacitance per unit of length of the wire.

16. $u = \frac{16A}{\pi^2}\sum_{n,m=0}^{\infty}\frac{\sin[(2n+1)x]\sin[(2m+1)y]}{(2n+1)^3(2m+1)^3}\cos(\pi a \lambda_{n,m}t)$, where

$$\lambda_{n,m} = \frac{1}{\pi}\sqrt{(2n+1)^2 + (2m+1)^2}.$$

17. $u = [\sin t - \sin(t\sqrt{2})]\sin x \sin y.$

18. $u = \sum_{n=1}^{\infty} A_n \cos\left[c\pi\sqrt{\left(\frac{n}{a}\right)^2 + \frac{1}{b^2}}\,t\right]\sin\left(\frac{n\pi}{a}x\right)\sin\left(\frac{\pi}{b}b\right) +$

$$+\sin\left(\frac{\pi}{b}y\right)\frac{\sinh[\frac{\pi}{b}(a-x)]}{\sinh\left(\frac{\pi a}{b}\right)},$$

where

$$A_n = \frac{2}{a\sinh\left(\frac{\pi a}{b}\right)}\int_0^a \sinh\left[\frac{\pi}{b}(x-a)\right]\sin\left(\frac{n\pi}{a}x\right)dx.$$

19. $u = \frac{P}{T}\left[\frac{1}{4}(b^2 - \rho^2) - 2b^2\sum_{k=1}^{\infty}\frac{J_0\left(\frac{\mu_k}{b}\rho\right)}{\mu_k^3 J_1(\mu_k)}\cos\left(\frac{\mu_k a}{b}t\right)\right]$,

where μ_k is the kth positive root of the Bessel function $J_0(x)$.

20. $u = J_0\left(\frac{\mu_1}{a}\rho\right)\frac{a^2}{\mu_1^2}\left[1 - \cos\left(\frac{\mu_1}{a}t\right)\right].$

21. $u = \cos\left(\sqrt{\left(\frac{\mu_2}{a}\right)^2 + \left(\frac{3\pi}{b}\right)^2}\,t\right)J_0\left(\frac{\mu_2}{a}\rho\right)\cos\left(\frac{3\pi}{b}z\right).$

22. $u = \left[\frac{1}{25}t - \frac{1}{125}\sin(5t)\right]\cos(3x)\cos(4y).$

23. $u = \frac{l^2}{4\pi^2}\sin\left[\left(\frac{2\pi}{l}\right)^2 t\right]\sin\left(\frac{2\pi}{l}x\right).$

24. $u = \cos x\cos(2t) + \frac{\varepsilon}{64}t\sin(2t)\cos x.$

25. First multiply both sides of the equation of free oscillations of a membrane by $\partial u/\partial t$ and then use the integration-by-parts formula

$$\int_{\Omega} \frac{\partial u}{\partial x}(x,y)v(x,y)\,dx\,dy = \int_{\partial \Omega} uv\cos(n,x)ds - \int_{\Omega} u(x,y)\frac{\partial v}{\partial x}(x,y)\,dx\,dy,$$

where Ω is a bounded domain in the plane and $\vec{n}$ is the unit outward normal to the boundary $\partial \Omega$ of Ω.

26. $u = \dfrac{a}{c\mu_5}J_3\left(\dfrac{\mu_5}{a}\rho\right)\cos(3\varphi)\sin\left(\dfrac{c\mu_5}{a}t\right).$

27. $u = \cos\varphi \displaystyle\sum_{m=1}^{\infty} A_m J_1\left(\dfrac{\mu_m}{a}\rho\right)\cos\left(\dfrac{\mu_m}{a}t\right) + A\dfrac{\rho}{a}\cos\varphi$, where μ_m denotes the mth positive root of the Bessel function $J_1(x)$ and the coefficients of the series are given by

$$A_m = \frac{-A\int_0^a \rho^2 J_1\left(\frac{\mu_m}{a}\rho\right)d\rho}{\frac{a^3}{2}[J_1'(\mu_m)]^2}.$$

28. $u = A\dfrac{J_0\left(\frac{\omega}{c}\rho\right)}{J_0\left(\frac{\omega}{c}a\right)}\cos(\omega t) - \displaystyle\sum_{n=1}^{\infty} A_n J_0\left(\dfrac{\mu_n}{a}\rho\right)\cos\left(\dfrac{c\mu_n}{a}t\right)$, where

$$A_n = \frac{2A\int_0^a \rho J_0\left(\frac{\omega}{c}\rho\right)J_0\left(\frac{\mu_n}{a}\rho\right)d\rho}{a^2 J_0\left(\frac{\omega}{c}a\right)[J_1(\mu_n)]^2},$$

c is the constant in the equation $u_{tt} = c^2 \Delta u$ and $\mu_n > 0$ are the roots of the equation $J_0(x) = 0$.

29. $u = \sin(2\varphi)\displaystyle\sum_{m=1}^{\infty} A_n J_2\left(\dfrac{\mu_m}{a}\rho\right)\cos\left(\dfrac{c\mu_m}{a}t\right) + A\left(\dfrac{\rho}{a}\right)^2\sin(2\varphi)$, where

$$A_m = -\frac{2A\int_0^a \rho^3 J_2\left(\frac{\mu_m}{a}\rho\right)d\rho}{a^4[J_2'(\mu_m)]^2}$$

and $\mu_m > 0$ are the roots of the equation $J_2(x) = 0$.

30. $u = A\dfrac{J_0\left(\frac{\omega}{c}\rho\right)}{J_0\left(\frac{\omega}{c}\rho\right)}\sin(\omega t) - \displaystyle\sum_{n=1}^{\infty} A_n J_0\left(\dfrac{\mu_n}{a}\rho\right)\sin\left(\dfrac{c\mu_n}{a}t\right)$, where

$$A_n = \frac{2A\omega}{c\mu_n a J_0\left(\frac{\omega}{c}a\right)[J_1(\mu_n)]^2}\int_0^a \rho J_0\left(\frac{\omega}{c}\rho\right)J_0\left(\frac{\mu_n}{a}\rho\right)d\rho,$$

c is the constant from the equation $u_{tt} = c^2 \Delta u$ and $\mu_n > 0$ are the roots of the equation $J_0(x)$.

31. $u = \dfrac{1}{c^2\left[\left(\frac{\mu_1}{a}\right)^2 + \frac{4\pi^2}{h^2}\right]} \left[t - \dfrac{\sin\left(c\sqrt{\left(\frac{\mu_1}{a}\right)^2 + \frac{4\pi^2}{h^2}}\,t\right)}{c\sqrt{\left(\frac{\mu_1}{a}\right)^2 + \frac{4\pi^2}{h^2}}}\right] \times$

$$\times J_0\left(\frac{\mu_1}{a}\rho\right)\sin\left(\frac{2\pi}{h}z\right).$$

32. $u = \dfrac{\rho^n P_n(\cos\theta)}{2n\rho_0^{n-1}}\,t^2 - \dfrac{2P_n(\cos\theta)}{a^2 n\rho_0^{n-3}}\sum_{k=1}^{\infty}\dfrac{1}{(\mu_k^{(n)})^2}\times$

$$\times \sin^2\left(\frac{a\mu_k^{(n)}t}{2\rho_0}\right)\frac{J_{n+1/2}\left(\frac{\mu_k^{(k)}}{\rho_0}\rho\right)}{\sqrt{\rho}},$$

where the coefficients A_k are given by the formula

$$A_k = \frac{1}{\frac{\rho_0^2}{2}\left[1 - \frac{n(n+1)}{(\mu_k^{(n)})^2}\right]J_{n+1/2}^2(\mu_k^{(n)})}\int_0^{\rho_0}\rho^{n+3/2}J_{n+1/2}\left(\frac{\mu_k^{(n)}}{\rho_0}\rho\right)d\rho,$$

ρ_0 is the radius of the container, $\mu_k^{(n)}$ are the positive roots of the equation

$$xJ_{n+1/2}'(x) - \frac{1}{2}J_{n+1/2}(x) = 0$$

and a is the constant in the equation $u_{tt} = a^2\Delta u$.

33. *Hint.* Consider the two problems

$$\begin{cases} u_{tt} = a^2\Delta u, \\ u|_{t=0} = u_0(x), \quad u_t|_{t=0} = u_1(x) \end{cases}$$

and

$$\begin{cases} u_{tt} = a^2\Delta u + f(x)g(t), \\ u|_{t=0} = u_0(x), \quad u_t|_{t=0} = 0. \end{cases}$$

34. $u = A\dfrac{J_n\left(\frac{\omega}{c}\rho\right)}{J_n\left(\frac{\omega}{c}a\right)}\sin(\omega t)\cos(n\varphi) - \cos(n\varphi)\sum_{m=1}^{\infty}B_m J_n\left(\frac{\mu_m}{a}\rho\right)\sin\left(c\frac{\mu_m}{a}\rho\right),$

where $J_n(x)$ is the Bessel function of first kind of index n, μ_m is the mth positive root of the equations $J_n(x) = 0$ and c is the constant in the equation $u_{tt} = c^2\Delta u$. The coefficients B_m are given by the formula

$$B_m = \frac{2\omega A}{caJ_n\left(\frac{\omega}{c}a\right)[J_n'(\mu_m)]^2\mu_m}\int_0^a \rho J_n\left(\frac{\omega}{c}\rho\right)J_n\left(\frac{\mu_m}{a}\rho\right)d\rho.$$

35. *Hint.* The assertion is established by differentiating with respect to the variable t the equation

$$\frac{\partial^2 u_f}{\partial t^2} = a^2 \Delta u_f.$$

36. $u = \ln \left(1 + t \int_{|y|=1} f(x + aty) d\omega_y\right),$

where $x = (x_1, x_2, x_3)$ and $y = (y_1, y_2, y_3).$

37. *Hint.* Let $u(t, x)$ be the solution of the Cauchy problem

$$\begin{cases} u_{tt} + \Delta^2 u = 0, \\ u|_{t=0} = f(x), \quad u_t|_{t=0} = 0. \end{cases}$$

Then the function $v(x, t) = u(x, t) + i \int_0^t \Delta u(x, \tau) d\tau$ is a solution of the Cauchy problem

$$\begin{cases} v_t = i\Delta v, \\ v|_{t=0} = f(x). \end{cases} \tag{$*$}$$

Indeed, we have

$$v_t = u_t + i\Delta u(x, t), \qquad v_{tt} = u_{tt} + i\Delta u_t(x, t)$$

and

$$i\Delta v = i\Delta u - \int_0^t \Delta^2 u(x, \tau) d\tau = i\Delta u + \int_0^t u_{tt}(x, \tau) d\tau =$$

$$= i\Delta u + u_t(x, t) - u_t(x, 0) = i\Delta u + u_t.$$

Therefore,

$$v_t = u_t + i\Delta u \qquad \text{and} \qquad i\Delta v = u_t + i\Delta u,$$

which proves that v solves $(*)$.
The converse assertion is proved in much the same way.

38. $v = \operatorname{Re} u.$
Hint. If $u = u_1 + iu_2$, then $(u_1 + iu_2)_t = i\Delta(u_1 + iu_2)$, whence $u_{1t} = -\Delta u_2$ and $u_{2t} = \Delta u_1$. Further differentiation yields $u_{1tt} = -\Delta u_{2t}$ and $\Delta u_{2t} = \Delta^2 u_{1t}$. Therefore, $u_1 = \operatorname{Re} u$ satisfies the equation

$$u_{1tt} + \Delta^2 u_1 = 0,$$

and the conditions $u_1|_{t=0} = f(x)$, $u_{1t}|_{t=0} = 0$ (indeed, from $u_2|_{t=0} = 0$ it follows that $\Delta u_2|_{t=0} = 0$, i.e., $u_{1t}|_{t=0} = 0$).

39. $u = \displaystyle\int_0^t (t - \tau) f(x, \tau) d\tau.$

Hint. Notice that when the function f does not depend on t, then the solution of the Cauchy problem

$$\begin{cases} u_{tt} + \Delta^2 u = 0, \\ u|_{t=0} = 0, \quad u_t|_{t=0} = f(x) \end{cases}$$

is the function $u(t, x) = t f(x)$. It follows that in order to solve our problem we need to use Duhamel's method.

40. $u = u_0 + t u_1.$

Chapter 3
Parabolic problems

There are many nonstationary phenomena that are described by parabolic, rather than hyperbolic, equations. Second-order partial differential equations of parabolic type are often encountered in the study of heat conduction and diffusion processes. The simplest equation of parabolic type,

$$u_t = a^2 u_{xx},$$

(where $a^2 = \text{const}$) is usually called the *heat equation*; here $u(x,t)$ is the temperature in the point with abscissa x at time t.

This equations says that the temperature in the point x and at time t will increase ($u_t > 0$) or decrease ($u_t < 0$) according to whether the second derivative u_{xx} is positive or negative (Figures 3.1 and 3.2; the arrows indicate the direction in which the temperature changes – heat flows from the warmer to the colder parts of a rod of length l; the moment of time is fixed).

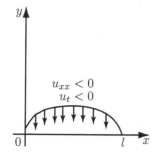

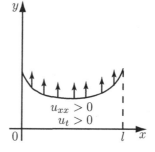

FIGURE 3.1.   FIGURE 3.2

Let us mention that several problems of electrodynamics also reduce to equations of parabolic type.

3.1. The Fourier integral transform method

We have already seen how the Fourier transformation can be used to solve hyperbolic problems. In this section we consider the application of the Fourier transformation to solving parabolic problems.

Problems on the line and half-line.

Example 1 [4, Ch. III, no. 55]. Applying the Fourier integral transformation, solve the following boundary value problem:

$$\begin{cases} u_t = a^2 u_{xx} + f(x,t), & -\infty < x < \infty, \quad t > 0, \\ u(x,0) = 0, & -\infty < x < \infty. \end{cases}$$

Solution. Denote by $\tilde{u}(\xi, t)$ the Fourier transform (with respect to the variable x) of the function $u(x, t)$, i.e.,

$$\tilde{u}(\xi, t) = \frac{1}{\sqrt{2\pi}} \int_{-\infty}^{\infty} u(x)e^{-i\xi x} dx,$$

or, in a simpler form, $\tilde{u}(\xi, t) = F[u(x, t)]$, where F is the Fourier transformation operator. It is well known that

$$F[u_{xx}(x, t)] = (i\xi)^2 F[u(x, t)],$$

i.e.,

$$F[u_{xx}(x, t)] = -\xi^2 \tilde{u}(\xi, t).$$

Further,

$$F[u_t(x, t)] = \tilde{u}_t(\xi, t), \qquad F[u(x, 0)] = \tilde{u}(\xi, 0) = 0.$$

Also, consider $F[f(x, t)] = \tilde{f}(\xi, t)$. Then the heat equation in Fourier transforms reads

$$\tilde{u}_t(\xi, t) + \xi^2 a^2 \tilde{u}(\xi, t) = \tilde{f}(\xi, t).$$

This is a first-order ordinary differential equation in the variable t (with ξ playing the role of a parameter). Therefore, we have the following Cauchy problem:

$$\begin{cases} \tilde{u}_t(\xi, t) + \xi^2 a^2 \tilde{u}(\xi, t) = \tilde{f}(\xi, t), & t > 0, \\ \tilde{u}(\xi, 0) = 0. \end{cases}$$

We will solve this Cauchy problem using Duhamel's method. To this end let us solve first the auxiliary problem

$$\begin{cases} v_t(\xi, t) + \xi^2 a^2 v(\xi, t) = 0, & t > 0, \\ v(\xi, t)|_{t=\tau} = \tilde{f}(\xi, \tau). \end{cases}$$

We have

$$v(\xi, t; \tau) = \tilde{f}(\xi, \tau)e^{-\xi^2 a^2 (t-\tau)}.$$

Then, as we know,

$$\tilde{u}(\xi, t) = \int_0^t v(\xi, t; \tau) \, d\tau = \int_0^t \tilde{f}(\xi, \tau)e^{-\xi^2 a^2 (t-\tau)} d\tau.$$

It follows that

$$u(x, t) = \frac{1}{\sqrt{2\pi}} \int_{-\infty}^{\infty} \tilde{u}(\xi, t)e^{i\xi x} d\xi = \frac{1}{\sqrt{2\pi}} \int_0^t d\tau \int_{-\infty}^{\infty} e^{-\xi^2 a^2 (t-\tau)} \tilde{f}(\xi, \tau)e^{i\xi x} d\xi =$$

$$= \frac{1}{\sqrt{2\pi}} \int_0^t d\tau \int_{-\infty}^{\infty} e^{-\xi^2 a^2 (t-\tau)} e^{i\xi x} d\xi \int_{-\infty}^{\infty} \frac{1}{\sqrt{2\pi}} e^{-i\lambda\xi} f(\lambda, \tau) d\lambda =$$

$$= \frac{1}{2\pi} \int_0^\tau d\tau \int_{-\infty}^{\infty} f(\lambda, \tau) d\lambda \int_{-\infty}^{\infty} e^{-\xi^2 a^2 (t-\tau)} e^{-i\xi(\lambda-x)} d\xi =$$

$$= \frac{1}{2\pi} \int_0^\tau d\tau \int_{-\infty}^{\infty} f(\lambda, \tau) d\lambda \cdot e^{-\frac{(\lambda-x)^2}{4a^2(t-\tau)}} \sqrt{\frac{\pi}{a^2(t-\tau)}} =$$

$$= \frac{1}{2a\sqrt{\pi}} \int_0^t d\tau \int_{-\infty}^{\infty} f(\lambda, \tau) \frac{e^{-\frac{(\lambda-x)^2}{4a^2(t-\tau)}}}{\sqrt{t-\tau}} d\lambda.$$

In this calculation we have used the well-known expression for the Fourier transform of the function e^{-bx^2} $(b > 0)$, namely

$$F[e^{-bx^2}] = \frac{1}{\sqrt{2\pi}} \int_{-\infty}^{\infty} e^{-bx^2} e^{-i\lambda x} dx = \frac{1}{\sqrt{2b}} e^{-\frac{\lambda^2}{4b}}.$$

We conclude that the solution of our problem is

$$u(x,t) = \frac{1}{2a\sqrt{\pi}} \int_0^t d\tau \int_{-\infty}^{\infty} f(\lambda, \tau) \frac{e^{-\frac{(\lambda-x)^2}{4a^2(t-\tau)}}}{\sqrt{t-\tau}} d\lambda.$$

Example 2 [4, Ch. III, no. 57]. Applying the Fourier integral transformation, solve the boundary value problem

$$\begin{cases} u_t = a^2 u_{xx}, & 0 < x, t < \infty, \\ u_x(0,t) = 0, & 0 < t < \infty, \\ u(x,0) = f(x), & 0 < x < \infty. \end{cases}$$

Solution. Let us apply the Fourier cosine transformation

$$\tilde{u}^c(\lambda, t) = \sqrt{\frac{2}{\pi}} \int_0^\infty u(x,t) \cos(\lambda x) \, dx.$$

Recall that the inverse Fourier cosine transformation is given by the rule

$$u(x,t) = \sqrt{\frac{2}{\pi}} \int_0^\infty \tilde{u}^c(\lambda, t) \cos(\lambda x) \, d\lambda.$$

From this equality it readily follows that

$$u_x(x,t) = \sqrt{\frac{2}{\pi}} (-\lambda) \int_0^\infty \tilde{u}^c(\lambda, t) \sin(\lambda x) \, d\lambda,$$

whence $u_x(0,t) = 0$, i.e., the boundary condition in our problem is satisfied. Further, observe that

$$\tilde{u}_{xx}^c(\lambda, t) = (-\lambda)^2 \sqrt{\frac{2}{\pi}} \int_0^\infty u(x,t) \cos(\lambda x)\, dx$$

and

$$\tilde{f}^c(\lambda) = \sqrt{\frac{2}{\pi}} \int_0^\infty f(x) \cos(\lambda x)\, dx.$$

In this way we obtain the following Cauchy problem for an ordinary differential equation:

$$\begin{cases} \tilde{u}_t^c(\lambda, t) + a^2 \lambda^2 \tilde{u}^c(\lambda, t) = 0, & t > 0, \\ \tilde{u}^c(\lambda, 0) = \tilde{f}^c(\lambda), \end{cases}$$

where λ is regarded as a parameter.

The solution of this problem has the form

$$\tilde{u}^c(\lambda, t) = \tilde{f}^c(\lambda) e^{-a^2 \lambda^2 t}.$$

Applying the inverse Fourier cosine transformation we obtain

$$u(x,t) = \sqrt{\frac{2}{\pi}} \int_0^\infty \tilde{f}^c(\lambda) e^{-a^2 \lambda^2 t} \cos(\lambda x)\, d\lambda =$$

$$= \frac{2}{\pi} \int_0^\infty \left(\int_0^\infty f(\xi) \cos(\lambda \xi)\, d\xi \right) \cos(\lambda x) e^{-a^2 \lambda^2 t} d\lambda =$$

$$= \frac{1}{\pi} \int_0^\infty f(\xi)\, d\xi \int_0^\infty e^{-a^2 \lambda^2 t} [\cos \lambda(x + \xi) + \cos \lambda(x - \xi)]\, dx =$$

$$= \frac{1}{\pi} \int_0^\infty f(\xi)\, d\xi \left[\int_0^\infty e^{-a^2 \lambda^2 t} \cos \lambda(x + \xi)\, d\lambda + \int_0^\infty e^{-a^2 \lambda^2 t} \cos \lambda(x - \xi)\, d\lambda \right] =$$

$$= \frac{1}{2a\sqrt{\pi t}} \int_0^\infty f(\xi) \left[e^{-\frac{(x+\xi)^2}{4a^2 t}} + e^{-\frac{(x-\xi)^2}{4a^2 t}} \right] d\xi.$$

In this calculation we used the fact that

$$\int_0^\infty e^{-a^2 y^2} \cos(\beta y)\, dy = \frac{\sqrt{\pi}}{2a} e^{-\frac{\beta^2}{4a^2}} \qquad (\alpha, \beta > 0).$$

Example 3 [4, Ch. III, no. 113]. Find the temperature distribution along an infinite rod composed of two homogeneous rods in contact at the point $x = 0$, with characteristics a_1, k_1 and a_2, k_2, respectively (here a_1, a_2 are the constants in the heat

equation and k_1, k_2 are the heat conductivity coefficients). The initial temperature
is

$$u(x,0) = \begin{cases} T_1, & \text{if } x < 0, \\ T_2, & \text{if } x > 0. \end{cases}$$

Solution. From the Poisson formula

$$u(x,t) = \frac{1}{2a\sqrt{\pi t}} \int_{-\infty}^{\infty} u(\xi,0) e^{-\frac{(x-\xi)^2}{4a^2 t}} \, d\xi,$$

which gives the temperature distribution along an infinite rod under the initial
condition $u(x,0)$, it follows that the solution of the heat equation

$$u_t = a^2 u_{xx}, \qquad -\infty < x < \infty, \quad t > 0,$$

with initial condition

$$u(x,0) = \begin{cases} T^*, & \text{if } x > 0, \\ T_2, & \text{if } x < 0 \end{cases}$$

is given by the function

$$u(x,t) = \frac{T + T^*}{2} + \frac{T^* - T}{2} \Phi\left(\frac{x}{2a\sqrt{t}}\right),$$

where the function $\Phi(x) = \frac{2}{\sqrt{\pi}} \int_0^x e^{-y^2} \, dy$ is the so-called *error integral*.

To solve the original problem we will argue as follows. First we extend the
left rod (with temperature T_1) to the right to obtain an infinite homogeneous rod.
Next we find the temperature of the resulting infinite rod under the assumption
that its initial temperature is

$$u_1(x,0) = \begin{cases} T_1, & \text{if } x < 0, \\ T_1^*, & \text{if } x > 0, \end{cases}$$

where T_1^* is some constant.

We then proceed in a similar manner with the right semi-infinite rod, assuming that

$$u_2(x,0) = \begin{cases} T_2, & \text{if } x > 0, \\ T_2^*, & \text{if } x < 0. \end{cases}$$

The constants T_1^* and T_2^* are found from the conjugation (matching) conditions:

(1) continuity of the temperature at the point $x = 0$:

$$u_1(0,t) = u_2(0,t);$$

(2) continuity of the heat flux at the point $x = 0$:

$$k_1 \frac{\partial u_1}{\partial x}(0,t) = k_2 \frac{\partial u_2}{\partial x}(0,t).$$

From the formula given above it follows that

$$u_1(x,t) = \frac{T_1 + T_1^*}{2} + \frac{T_1^* - T_1}{2}\Phi\left(\frac{x}{2a_1\sqrt{t}}\right),$$

$$u_2(x,t) == \frac{T_2 + T_2^*}{2} + \frac{T_2 - T_2^*}{2}\Phi\left(\frac{x}{2a_2\sqrt{t}}\right).$$

Hence, the conjugation conditions yield the following system of equations with unknowns T_1^* and T_2^*:

$$\begin{cases} T_1^* - T_2^* = T_2 - T_1, \\ \dfrac{k_1}{a_1}(T_1^* - T_1) = \dfrac{k_2}{a_2}(T_2 - T_2^*). \end{cases}$$

Solving this system, we obtain

$$T_1^* = \frac{2\frac{k_2}{a_2}T_2 + \left(\frac{k_1}{a_1} - \frac{k_2}{a_2}\right)T_1}{\frac{k_1}{a_1} + \frac{k_2}{a_2}},$$

$$T_2^* = \frac{2\frac{k_1}{a_1}T_1 + \left(\frac{k_2}{a_2} - \frac{k_1}{a_1}\right)T_2}{\frac{k_1}{a_1} + \frac{k_2}{a_2}}.$$

Therefore,

$$u_1(x,t) = T_0 + (T_0 - T_1)\Phi\left(\frac{-x}{2a_1\sqrt{t}}\right), \qquad -\infty < x < 0,$$

$$u_2(x,t) = T_0 + (T_0 - T_2)\Phi\left(\frac{x}{2a_2\sqrt{t}}\right), \qquad 0 < x < \infty,$$

where

$$T_0 = \frac{\frac{k_1}{a_1}T_1 + \frac{k_2}{a_2}T_2}{\frac{k_1}{a_1} + \frac{k_2}{a_2}}.$$

Problems in three-dimensional space and half-space.

Example 4 [4, Ch.V, no. 66]. Solve the Cauchy problem

$$\begin{cases} u_t = a^2(u_{xx} + u_{yy} + u_{zz}) + g(x,y,z,t), & -\infty < x,y,z < \infty, \quad t > 0, \\ u|_{t=0} = 0, & -\infty < x,y,z < \infty. \end{cases}$$

Solution. We use the Fourier transformation with respect to the space variables:

$$\tilde{u}(\lambda,\mu,\nu,t) = \frac{1}{(2\pi)^{3/2}}\iiint\limits_{-\infty}^{\infty} u(x,y,z,t)e^{-i(\lambda x + \mu y + \nu z)}\,dx\,dy\,dz,$$

$$\widetilde{g}(\lambda, \mu, \nu, t) = \frac{1}{(2\pi)^{3/2}} \iiint\limits_{-\infty}^{\infty} g(x, y, z, t) e^{-i(\lambda x + \mu y + \nu z)} \, dx \, dy \, dz.$$

We obtain a Cauchy problem for the Fourier transforms:

$$\begin{cases} \widetilde{u}_t + a^2(\lambda^2 + \mu^2 + \nu^2)\widetilde{u} = \widetilde{g}(\lambda, \mu, \nu, t), & t > 0, \\ \widetilde{u}|_{t=0} = 0. \end{cases}$$

Applying here Duhamel's method we have

$$\widetilde{u}(\lambda, \mu, \nu, t) = \int_0^t \widetilde{g}(\lambda, \mu, \nu, t) e^{-(\lambda^2 + \mu^2 + \nu^2)(t-\tau)} d\tau.$$

Now using the inverse Fourier transformation we obtain

$$u(x, y, x, t) = \frac{1}{(2\pi)^{3/2}} \iiint\limits_{-\infty}^{\infty} \widetilde{u}(\lambda, \mu, \nu, t) e^{i(\lambda x + \mu y + \nu z)} \, d\lambda \, d\mu \, d\nu =$$

$$= \frac{1}{(2\pi)^3} \iiint\limits_{-\infty}^{\infty} \left[\int_0^t \left(\iiint\limits_{-\infty}^{\infty} g(\xi, \eta, \zeta, \tau) \, e^{-i(\lambda\xi + \mu\eta + \nu\zeta)} \, d\xi \, d\eta \, d\zeta \right) e^{-(\lambda^2 + \mu^2 + \nu^2)(t-\tau)} d\tau \right] \times$$

$$\times e^{i(\lambda x + \mu y + \nu z)} \, d\lambda \, d\mu \, d\nu =$$

$$= \frac{1}{(2\pi)^3} \int_0^t d\tau \iiint\limits_{-\infty}^{\infty} g(\xi, \eta, \zeta, \tau) \, d\xi \, d\eta \, d\zeta \times$$

$$\times \iiint\limits_{-\infty}^{\infty} e^{-(\lambda^2 + \mu^2 + \nu^2)(t-\tau)} e^{i[\lambda(x-\xi) + \mu(y-\eta) + \nu(z-\zeta)]} \, d\lambda \, d\mu \, d\nu.$$

But

$$\iiint\limits_{-\infty}^{\infty} e^{-a^2(\lambda^2 + \mu^2 + \nu^2)(t-\tau) + i[\lambda(x-\xi) + \mu(y-\eta) + \nu(z-\zeta)]} \, d\lambda \, d\mu \, d\nu =$$

$$= \frac{\pi^{3/2}}{a^3(t-\tau)^{3/2}} e^{-\frac{(x-\xi)^2 + (y-\eta)^2 + (z-\zeta)^2}{4a^2 t}}.$$

We conclude that

$$u(x, y, z, t) = \frac{1}{(2a\sqrt{\pi})^3} \int_0^t \frac{d\tau}{(t-\tau)^{3/2}} \iiint\limits_{-\infty}^{\infty} e^{-\frac{(x-\xi)^2 + (y-\eta)^2 + (z-\zeta)^2}{4a^2(t-\tau)}} g(\xi, \eta, \zeta, \tau) d\xi \, d\eta \, d\zeta.$$

Remark 1. Suppose that $g(x, y, z, t)$ does not depend on t, i.e., we are dealing with the Cauchy problem

$$\begin{cases} u_t = a^2(u_{xx} + u_{yy} + u_{zz}) + g(x, y, z), & -\infty < x, y, z < \infty, \quad t > 0, \\ u|_{t=0} = 0, & -\infty < x, y, z < \infty. \end{cases}$$

Then the solution is given by

$$u(x, y, z, t) = \frac{1}{4\pi a^2} \iiint\limits_{-\infty}^{\infty} \frac{g(\xi, \eta, \zeta)}{r} \left[1 - \Phi\left(\frac{r}{2a\sqrt{t}} \right) \right] d\xi \, d\eta \, d\zeta,$$

where

$$\Phi(\alpha) = \frac{2}{\sqrt{\pi}} \int_0^\alpha e^{-p^2} dp, \qquad r = \sqrt{(x - \xi)^2 + (y - \eta)^2 + (z - \zeta)^2}.$$

Indeed, let us denote $\omega = \frac{r}{2a\sqrt{t-\tau}}$. Then

$$u(x, y, z, t) = \frac{1}{(2a\sqrt{\pi})^3} \int_0^t \frac{d\tau}{(t-\tau)^{3/2}} \iiint\limits_{-\infty}^{\infty} e^{-\frac{(x-\xi)^2+(y-\eta)^2+(z-\zeta)^2}{4a^2(t-\tau)}} g(\xi, \eta, \zeta) d\xi \, d\eta \, d\zeta =$$

$$= \frac{1}{4\pi a^2} \iiint\limits_{-\infty}^{\infty} g(\xi, \eta, \zeta) \left(\frac{1}{2a\sqrt{\pi}} \frac{4a}{r} \int_{\frac{r}{2a\sqrt{t}}}^{\infty} e^{-\omega^2} d\omega \right) d\xi \, d\eta \, d\zeta =$$

$$= \frac{1}{4\pi a^2} \iiint\limits_{-\infty}^{\infty} \frac{g(\xi, \eta, \zeta)}{r} \left[\frac{2}{\sqrt{\pi}} \left(\frac{\sqrt{\pi}}{2} - \int_0^{\frac{r}{2a\sqrt{t}}} e^{-\omega^2} d\omega \right) \right] d\xi \, d\eta \, d\zeta =$$

$$= \frac{1}{4\pi a^2} \iiint\limits_{-\infty}^{\infty} \frac{g(\xi, \eta, \zeta)}{r} \left[1 - \Phi\left(\frac{r}{2a\sqrt{t}} \right) \right] d\xi \, d\eta \, d\zeta.$$

Remark 2. Suppose $g(x, y, z, t)$ does not depend on z, i.e., we are dealing with the Cauchy problem

$$\begin{cases} u_t = a^2(u_{xx} + u_{yy}) + g(x, y, t), & -\infty < x, y < \infty, \quad t > 0, \\ u|_{t=0} = 0, & \infty < x, y < \infty. \end{cases}$$

Then clearly in this case the solution is given by

$$u(x, y, t) = \frac{1}{4\pi a^2} \int_0^t \frac{d\tau}{t - \tau} \iint\limits_{-\infty}^{\infty} g(\xi, \eta, \zeta) e^{-\frac{(x-\xi)^2+(y-\eta)^2}{4a^2(t-\tau)}} d\xi \, d\eta.$$

Example 5 [4, Ch. V, no. 67]. Find the temperature distribution in the upper half-space if the initial temperature in the half-space is zero, the temperature in the plane $z = 0$ is zero, and in the half-plane itself there is a distribution of heat sources with density $f(x, y, z, t)$, i.e., solve the Cauchy problem

$$\begin{cases} u_t = a^2(u_{xx} + u_{yy} + u_{zz}) + f(x, y, z, t), & -\infty < x, y < \infty, \quad z > 0, \quad t > 0, \\ u|_{z=0} = 0, & -\infty < x, y < \infty, \quad t \geq 0, \\ u|_{t=0} = 0, & -\infty < x, y < \infty, \quad z \geq 0. \end{cases}$$

Solution. In the present case we will take the Fourier transformation in the form

$$\widetilde{u}(\lambda, \mu, \nu, t) = \frac{1}{2^{1/2}\pi^{3/2}} \iint\limits_{-\infty}^{\infty} \int_0^\infty u(x, y, z, t) e^{-i(\lambda x + \mu y)} \sin(\nu z) \, dx \, dy \, dz.$$

Then

$$u(z, y, z, t) = \frac{1}{2^{1/2}\pi^{3/2}} \iint\limits_{-\infty}^{\infty} \int_0^\infty \widetilde{u}(\lambda, \mu, \nu, t) e^{i(\lambda x + \mu y)} \sin(\nu z) \, d\lambda \, d\mu \, d\nu$$

(note that this guarantees that the condition $u|_{z=0} = 0$ is satisfied). The ordinary differential equation (with respect to t) for the Fourier transforms is

$$\begin{cases} \widetilde{u}_t + a^2(\lambda^2 + \mu^2 + \nu^2)\widetilde{u} = \widetilde{f}(\lambda, \mu, \nu, t), & t > 0, \\ u|_{t=0} = 0. \end{cases}$$

The solution of this problem is given by

$$\widetilde{u}(\lambda, \mu, \nu, t) = \int_0^t \widetilde{f}(\lambda, \mu, \nu, t) e^{-a^2(\lambda^2 + \mu^2 + \nu^2)(t-\tau)} \, d\tau.$$

Therefore,

$$u(z, y, z, t) = \frac{1}{2^{1/2}\pi^{3/2}} \iint\limits_{-\infty}^{\infty} \int_0^\infty \widetilde{u}(\lambda, \mu, \nu, t) e^{i(\lambda x + \mu y)} \sin(\nu z) \, d\lambda \, d\mu \, d\nu =$$

$$= \frac{1}{2^{1/2}\pi^{3/2}} \iint\limits_{-\infty}^{\infty} \int_0^\infty \left[\int_0^t \frac{1}{2^{1/2}\pi^{3/2}} \left(\iint\limits_{-\infty}^{\infty} \int_0^\infty f(\xi, \eta, \zeta, \tau) e^{-i(\lambda\xi + \mu\eta)} \sin(\nu\zeta) \, d\xi \, d\eta \, d\zeta \right) \times \right.$$

$$\left. \times e^{-a^2(\lambda^2 + \mu^2 + \nu^2)(t-\tau)} d\tau \right] e^{i(\lambda x + \mu y)} \sin(\nu z) \, d\lambda \, d\mu \, d\nu =$$

$$= \frac{1}{2\pi^3} \int_0^t d\tau \iint_{-\infty}^{\infty} \int_0^{\infty} f(\xi, \eta, \zeta, \tau) \, d\xi \, d\eta \, d\zeta \times$$

$$\times \iint_{-\infty}^{\infty} \int_0^{\infty} e^{-a^2(\lambda^2+\mu^2)(t-\tau)} e^{i[(x-\xi)\lambda+(y-\eta)\mu]} e^{-a^2\nu^2(t-\tau)} \sin(\nu\zeta) \sin(\nu z) \, d\lambda \, d\mu \, d\nu.$$

Let us compute the second triple integral:

$$\iint_{-\infty}^{\infty} e^{-a^2(\lambda^2+\mu^2)(t-\tau)} e^{i[(x-\xi)\lambda+(y-\eta)\mu]} \, d\lambda \, d\mu \int_0^{\infty} e^{-a^2\nu^2(t-\tau)} \sin(\nu\zeta) \sin(\nu z) \, d\nu =$$

$$= \frac{\pi}{a^2(t-\tau)} e^{-\frac{(x-\xi)^2+(y-\eta)^2}{4a^2(t-\tau)}} \times$$

$$\times \frac{1}{2} \left[\int_0^{\infty} e^{-a^2\nu^2(t-\tau)} \cos\nu(\zeta-z) \, d\nu - \int_0^{\infty} e^{-a^2\nu^2(t-\tau)} \cos\nu(\zeta+z) \, d\nu \right] =$$

$$= \frac{\pi}{2a^2(t-\tau)^2} e^{-\frac{(x-\xi)^2+(y-\eta)^2}{4a^2(t-\tau)}} \left[\frac{\sqrt{\pi}}{2a\sqrt{t-\tau}} e^{-\frac{(\zeta-z)^2}{4a^2(t-\tau)}} - \frac{\sqrt{\pi}}{2a\sqrt{t-\tau}} e^{-\frac{(\zeta+z)^2}{4a^2(t-\tau)}} \right] =$$

$$= \frac{\pi^{3/2}}{4a^3(t-\tau)^{3/2}} \left[e^{-\frac{(\xi-x)^2+(\eta-y)^2+(\zeta-z)^2}{4a^2(t-\tau)}} - e^{-\frac{(\xi-x)^2+(\eta-y)^2+(\zeta+z)^2}{4a^2(t-\tau)}} \right].$$

We conclude that the solution of the original problem is given by the formula

$$u(x, y, z, t) = \frac{1}{(2a\sqrt{\pi})^3} \int_0^t \frac{d\tau}{(t-\tau)^{3/2}} \int_0^{\infty} d\zeta \times$$

$$\times \iint_{-\infty}^{\infty} f(\xi, \eta, \zeta, t) \left[e^{-\frac{(x-\xi)^2+(y-\eta)^2+(\zeta-z)^2}{4a^2(t-\tau)}} - e^{-\frac{(x-\xi)^2+(y-\eta)^2(\zeta+z)^2}{4a^2(t-\tau)}} \right] d\xi \, d\eta.$$

3.2. The Laplace integral transform method

We have already made acquaintance with the application of the Laplace transformation to hyperbolic equations. In this section we will use the Laplace transformation to solve several parabolic equations. Let us note that the block diagram for solving problems by this method is identical to that in the hyperbolic case.

Example 1 [3, Ch. 5, no. 832]. Solve the boundary value problem

$$\begin{cases} u_t = u_{xx} + u - f(x), & 0 < x, t < \infty, \\ u(0, t) = t, & u_x(0, t) = 0, & 0 \le t < \infty. \end{cases} \tag{3.1}$$

Solution. We shall use the Laplace transformation with respect to the variable x. Using its properties, we can write

$$u(x,t) \doteqdot U(p,t),$$

$$u_t(x,t) \doteqdot U_t(p,t),$$

$$u_x(x,t) \doteqdot pU(p,t) - t,$$

$$u_{xx}(x,t) \doteqdot p^2 U(p,t) - pt, \qquad f(x) \doteqdot F(p).$$

Applying the Laplace transformation to both sides of equation (3.1), we get

$$L[u_t] = L[u_{xx}] + L[u] - L[f(x)],$$

or

$$U_t(p,t) = p^2 U(p,t) - pt + U(p,t) - F(p).$$

Thus, we obtained a first-order ordinary differential equation (in which p plays the role of a parameter) in the independent variable t. It can be recast as

$$U_t - (1 + p^2)U = -[F(p) + pt].$$

The general solution of this equation is given by the function

$$U(p,t) = Ce^{(1+p^2)t} + \frac{p}{(1+p^2)^2} + \frac{F(p)}{1+p^2} + \frac{p}{1+p^2}t.$$

Now observe that the constant C must be set equal to zero; otherwise, $U(p,t) \to \infty$ as $p \to \infty$, which means that a necessary condition for the existence of the Laplace transformation is violated. Therefore,

$$U(p,t) = \frac{p}{(1+p^2)^2} + \frac{F(p)}{1+p^2} + \frac{p}{1+p^2}t.$$

Now it remains to return to the original (i.e., apply the inverse Laplace transformation). We have

$$\frac{p}{1+p^2} \doteqdot \cos x,$$

and, by the convolution theorem,

$$\frac{p}{(1+p^2)^2} \doteqdot \frac{1}{2}x\sin x, \qquad \frac{F(p)}{1+p^2} \doteqdot \int_0^x f(t)\sin(x-t)\,dt.$$

We conclude that the solution of our problem is

$$u(x,t) = t\cos x + \frac{1}{2}x\sin x + \int_0^x f(t)\sin(x-t)\,dt.$$

Example 2 [3, Ch. 5, no. 837(c)]. The initial temperature of a thin homogeneous rod is equal to zero. Determine the temeperature $u(x,t)$ in the rod for $t > 0$ if the rod is semi-infinite $(0 < x < \infty)$ and $u(0,t) = \mu(t)$, where $\mu(t)$ is a given function.

Solution. Mathematically, the problem is formulated as follows:

$$\begin{cases} u_t = a^2 u_{xx}, & 0 < x, t < \infty, \\ u(x,0) = 0, & 0 \le x < \infty, \\ u(0,t) = \mu(t), & 0 \le t < \infty. \end{cases} \tag{3.2}$$

To solve this problem we will use the Laplace transformation in the variable t. Again, using the properties of the Laplace transformation we have

$$u(x,t) \doteq U(x,p), \qquad u_t(x,t) \doteq pU(x,p), \qquad u_{xx}(x,t) \doteq U_{xx}(x,p).$$

Equation (3.2) becomes

$$L[u_t] = a^2 L[u_{xx}],$$

or

$$pU(x,p) = a^2 U_{xx}(x,p),$$

or, finally,

$$U_{xx} - \frac{p}{a^2} U(x,p) = 0.$$

We obtained a second-order ordinary differential equation in the independent variable x, where p plays the role of a parameter. This equation is supplemented by two conditions. Indeed, from the boundary condition of problem (3.2) we know the function $U(x,p)$ for $x = 0$. Let us denote the Laplace transform of the function $\mu(t)$ by $M(p)$, i.e.,

$$\mu(t) \doteq M(p).$$

Further, from the physical meaning of the problem it is obvious that we must take $U(\infty, p) = 0$. Thus, we arrive at the boundary value problem

$$\begin{cases} U_{xx} - \dfrac{p}{a^2} U(x,p) = 0, & 0 < x < \infty, \\ U(0,p) = M(p), \quad U(\infty, p) = 0. \end{cases} \tag{3.3}$$

Its general solution is

$$U(x,p) = C_1 e^{\frac{\sqrt{p}}{a}x} + C_2 e^{-\frac{\sqrt{p}}{a}x},$$

where C_1 and C_2 are arbitrary constants.

Clearly, $C_1 = 0$, since otherwise $U(x,p) \to \infty$ as $x \to \infty$. Therefore,

$$U(x,p) = C_2 e^{-\frac{\sqrt{p}}{a}x},$$

Setting here $x = 0$, we have $U(0, p) = C_2$; by the boundary condition at $x = 0$ in (3.3), we must put $C_2 = M(p)$, and so

$$U(x, p) = M(p)e^{-\frac{\sqrt{p}}{a}x},$$

It remains to go back to the original $u(x, t)$, i.e., apply the inverse Laplace transformation. Since $M(p) \doteqdot \mu(t)$ and

$$e^{-\frac{\sqrt{p}}{a}x} \doteqdot \frac{x}{2a\sqrt{\pi}t^{3/2}}e^{-\frac{x^2}{4a^2t}},$$

the convolution theorem yields

$$u(x, t) = \frac{x}{2a\sqrt{\pi}} \int_0^t \frac{\mu(\tau)}{(t - \tau)^{3/2}}e^{-\frac{x^2}{4a^2(t-\tau)}}\,d\tau.$$

Problem 3. The initial voltage in a semi-infinite homogeneous wire $0 < x < \infty$ is equal to zero. The self-inductance and loss per unit of length of the wire are assumed to be negligibly small. Starting at time $t = 0$, a constant electric driving force E_0 is applied at the extremity $(x = 0)$ of the wire. Find the voltage in the wire.

Solution. Mathematically, the problem is formulated as follows:

$$\begin{cases} v_t = \dfrac{1}{CR}v_{xx}, & 0 < x, t < \infty, \\ v(x, 0) = 0, & 0 \leq x < \infty, \\ v(0, t) = E_0, & 0 \leq t < \infty. \end{cases} \tag{3.4}$$

Here C and R are the capacitance and resistance per unit of length of the wire and $v(t, x)$ is the voltage in the wire in the point with coordinate x and at time t.

Applying the Laplace transformation with respect to t to both sides of equation (3.4) we obtain

$$L[v_t] = \frac{1}{CR}L[v_{xx}].$$

Denoting $V(x, s) = \int_0^\infty e^{-st}v(x, t)dt$, where s is a real variable taking values in the interval $(0, \infty)$, and observing that

$$v(x, t) \doteqdot V(x, s), \qquad v_t(x, t) \doteqdot sV(x, s),$$

$$v_{xx} \doteqdot V_{xx}(x, s), \qquad E_0 \doteqdot \frac{E_0}{s},$$

we obtain the following boundary value problem for an ordinary differential equation in the variable x:

$$\begin{cases} V_{xx}(x, s) - sCRV(x, s) = 0, & 0 < x < \infty, \\ V(0, s) = \dfrac{E_0}{s}. \end{cases}$$

Here s plays the role of a parameter.

We obtained a second-order ordinary differential equation, but apparently with only one boundary condition. But we actually have a second boundary condition, namely, from physical considerations it follows that $V(x, s) \to 0$ as $x \to \infty$. So we have the required number of boundary conditions.

The general solution of our equation is

$$V(x, s) = C_1 e^{\sqrt{sCR}x} + C_2 e^{-\sqrt{sCR}x},$$

where C_1 and C_2 are arbitrary constants. We immediately observe that $C_1 = 0$ (otherwise, $V(x, s) \to \infty$ as $x \to \infty$). Therefore, $V(x, s) = C_2 e^{-\sqrt{sCR}x}$, and setting here $x = 0$ we obtain $V(0, s) = C_2$. But $V(0, s) = \frac{E_0}{s}$, so

$$V(x, s) = \frac{E_0}{s} e^{-\sqrt{sCR}x},$$

and now it remains to invert the Laplace transformation. Consulting Table 2.2, we conclude that the voltage is

$$v(x, t) = E_0 \left[1 - \Phi \left(\frac{\sqrt{CR}\, x}{2\sqrt{t}} \right) \right],$$

where $\Phi(z) = \frac{2}{\sqrt{\pi}} \int_0^z e^{-x^2} dx$ is the error integral.

3.3. The Fourier method (method of separation of variables)

We have already encountered this method in our study of hyperbolic problems. The block diagram of the method of separation of variables for solving parabolic problems is analogous to that of the Fourier method for hyperbolic equations.

The case of a segment.

Example 1. Solve the mixed problem

$$\begin{cases} u_t = u_{xx}, & 0 < x < \pi, \quad t > 0, \\ u(x, 0) = \cos(2x), & 0 \le x \le \pi, \\ u_x(0, t) = u_x(\pi, t) = 0, & t \ge 0. \end{cases}$$

Solution. Applying the method of separation of variables, we first seek the solution of the auxiliary problem

$$\begin{cases} u_t = u_{xx}, & 0 < x < \pi, \quad t > 0, \\ u_x(0, t) = u_x(\pi, t) = 0, & t \ge 0 \end{cases}$$

in the form $u(x, t) = X(x)T(t)$. In this way we obtain a Sturm-Liouville problem for the function $X(x)$,

$$\begin{cases} X'' + \lambda X = 0, & 0 < x < \pi, \\ X'(0) = X'(\pi) = 0, \end{cases}$$

which has the solutions $\lambda_n = n^2$, $X_n(x) = \cos(nx)$, $n = 0, 1, 2, \ldots$, and the equation

$$T' + \lambda T = 0$$

for the determination of $T(t)$. Its solutions corresponding to the values λ_n are

$$T_n(t) = C_n e^{-n^2 t}, \qquad n = 0, 1, 2, \ldots$$

Thus, the solution "atoms" for our problem are the functions

$$u_n(x, t) = T_n(t) X_n(x),$$

i.e.,

$$u_n(x, t) = e^{-n^2 t} \cos(nx).$$

Hence, the solution of the original problem is given by a series

$$u(x, t) = \sum_{n=0}^{\infty} C_n e^{-n^2 t} \cos(nx).$$

In the present case the boundary conditions yield $C_2 = 1$, $C_n = 0$, $n \neq 2$, so that

$$u(x, t) = e^{-4t} \cos(2x).$$

Problem 2. Solve the mixed problem

$$\begin{cases} u_t = u_{xx} + \sin t \sin(2x), & 0 < x < \pi, \quad t > 0, \\ u(x, 0) = 0, & 0 \le x \le \pi, \\ u(0, t) = 0, \quad u(\pi, t) = 0, & t \ge 0. \end{cases}$$

Solution. We will seek the solution $u(x, t)$ in the form $u(x, t) = f(t) \sin(2x)$. Inserting this expression in our equation we obtain

$$f' \sin(2x) = -4f \sin(2x) + \sin t \sin(2x),$$

whence

$$\begin{cases} f' + 4f = \sin t, & t \ge 0, \\ f(0) = 0. \end{cases}$$

Solving this Cauchy problem we obtain

$$f(t) = \frac{1}{17} e^{-4t} - \frac{1}{17} \cos t + \frac{4}{17} \sin t.$$

Therefore,

$$u(x, t) = \frac{\sin(2x)}{17} (e^{-4t} - \cos t + 4 \sin t).$$

The case of a circle.

Example 3. Solve the problem of heat propagation in a disc of radius b (Figure 3.3).

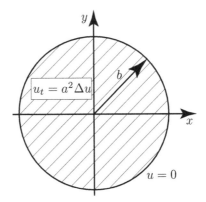

FIGURE 3.3. Heat propagation in a disc

The problem is formulated mathematically as follows:

$$\begin{cases} u_t = a^2(u_{xx} + u_{yy}^2), & 0 < \rho < b, \quad t > 0, \\ u|_{t=0} = J_0\left(\dfrac{\mu_{03}}{b}x\right), & 0 \le \rho \le b, \\ u|_{\rho=b} = 0, & t \ge 0, \end{cases} \tag{3.5}$$

where μ_{0n} is the nth positive root of the equation $J_0(x) = 0$.

Solution. Writing the Laplace operator $\Delta u = u_{xx} + u_{yy}$ in polar coordinates, we obtain the equation

$$u_t = a^2 \left[\frac{1}{\rho}\frac{\partial}{\partial \rho}\left(\rho\frac{\partial u}{\partial \rho}\right) + \frac{1}{\rho^2}\frac{\partial^2 u}{\partial \varphi^2}\right].$$

Since the initial and boundary conditions are independent of φ, the same is true about the solution. Hence, $u = u(\rho, t)$, i.e., we obtain the equation

$$u_t = a^2 \frac{1}{\rho}\frac{\partial}{\partial \rho}\left(\rho\frac{\partial u}{\partial \rho}\right).$$

Let us seek its solution in the form $u(\rho, t) = R(\rho)T(t)$, using also the condition $u(b, t) = 0$. Substituting this expression for u in the equation we obtain

$$RT' = a^2 \frac{1}{\rho}T\frac{d}{d\rho}(\rho R'),$$

or, separating the variables,

$$\frac{T'}{a^2 T} = \frac{\frac{1}{\rho} T \frac{d}{d\rho}(\rho R')}{R} = -\lambda.$$

This yields two ordinary differential equations:

$$(1) \qquad \frac{1}{\rho} \frac{d}{d\rho}(\rho R') + \lambda R = 0;$$

$$(2) \qquad T' + \lambda a^2 T = 0.$$

First consider equation (1), which can be recast as

$$R'' + \frac{1}{\rho} R' + \lambda R = 0,$$

or

$$\rho^2 R'' + \rho R' + \lambda \rho^2 R = 0.$$

Introducing the new variable $x = \sqrt{\lambda}\rho$ and denoting $y(x)|_{x=\sqrt{\lambda}\rho} = R(\rho)$, we obtain

$$R' = \sqrt{\lambda} y', \qquad R'' = \lambda y'',$$

where now the prime denotes differentiation with respect to x. Therefore, we have the equation

$$x^2 y'' + xy' + x^2 y = 0,$$

which is recognized to be the Bessel equation of index zero. Its solution is

$$y(x) = C_1 J_0(x) + C_2 Y_0(x),$$

where $J_0(x)$ and $Y_0(x)$ are the Bessel functions of index zero of the first and second kind, respectively, and C_1, C_2 are arbitrary constants. Clearly, $C_2 = 0$, because the solution of our problem must be finite and the center of the disc. Hence, $y(x) = C J_0(x)$, or, taking $C = 1$,

$$R(\rho) = J_0(\sqrt{\lambda}\rho).$$

By the boundary conditions, $R(b) = 0$, i.e.,

$$J_0(\sqrt{\lambda}b) = 0.$$

Recalling that the positive roots of the Bessel function J_0 are denoted by μ_{0n}, we find that the eigenvalues of our problem are $\lambda_n = \left(\frac{\mu_{0n}}{b}\right)^2$, where $n = 1, 2, \ldots$ The corresponding eigenfunctions are

$$R_n(\rho) = J_0\left(\frac{\mu_{0n}}{b}\rho\right), \qquad n = 1, 2, \ldots$$

Next, we find the corresponding solutions $T_n(t)$ of equation (2), namely,

$$T_n(t) = C_n e^{-\left(\frac{\mu_{0n}a}{b}\right)^2 t}.$$

Therefore, the solution the original problem is given by a series

$$u(\rho, t) = \sum_{n=1}^{\infty} C_n e^{-\left(\frac{\mu_{0n}a}{b}\right)^2 t} J_0\left(\frac{\mu_{0n}}{b}\rho\right).$$

Using the initial condition one readily sees that $C_3 = 1$ and $C_n = 0$ if $n \neq 3$. We conclude that the solution is

$$u(\rho, t) = e^{-\left(\frac{\mu_{03}a}{b}\right)^2 t} J_0\left(\frac{\mu_{03}}{b}\rho\right).$$

The case of a rectangle.

Example 4. Solve the following problem of heat propagation in a rectangle (Figure 3.4).

$$\begin{cases} u_t = a^2(u_{xx} + u_{yy}^2), & 0 < x < b_1, \quad 0 < y < b_2, \quad t > 0, & (3.6) \\ u|_{t=0} = \sin\left(\dfrac{2\pi x}{b_1}\right) \sin\left(\dfrac{\pi y}{b_2}\right), & 0 \le x \le b_1, \quad 0 \le y \le b_2, & (3.7) \\ u|_\Gamma = 0, & t \ge 0, & (3.8) \end{cases}$$

where Γ denotes the boundary of the rectangle.

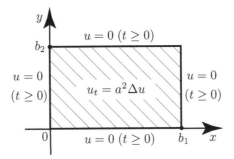

FIGURE 3.4. Heat propagation in a rectangle

We will seek the solution of problem (3.6), (3.8) in the form

$$u(x, y, t) = T(t)V(x, y).$$

Substituting this expression in the heat equation we obtain

$$T'V = a^2 T(V_{xx} + V_{yy}), \quad \text{or} \quad \frac{T'}{a^2 T} = \frac{V_{xx} + V_{yy}}{V} = -\lambda.$$

This yields the following two equations:

$$V_{xx} + V_{yy} + \lambda V = 0, \qquad (3.9)$$

$$T' + \lambda a^2 T = 0. \qquad (3.10)$$

Note that (3.9) is a partial differential equation, whereas (3.10) is an ordinary differential equation.

First let us solve (3.9). From the boundary condition $u|_\Gamma = 0$ is follows that $V|_\Gamma = 0$. Let us consider the eigenvalue problem

$$\begin{cases} V_{xx} + V_{yy} + \lambda V = 0, \\ V|_\Gamma = 0 \end{cases} \qquad (3.11)$$

and seek its solution in the form

$$V(x, y) = X(x)Y(y).$$

Substituting this expression in equation (3.11) we get

$$X''Y + XY'' + \lambda XY = 0,$$

or

$$\frac{X''}{X} + \frac{Y''}{Y} + \lambda = 0,$$

or, equivalently,

$$\frac{Y''}{Y} + \lambda = -\frac{X''}{X} = \mu.$$

This yields two ordinary differential equations:

$$X'' + \mu X = 0, \qquad (3.12)$$

$$Y'' + (\lambda - \mu)Y = 0. \qquad (3.13)$$

From the boundary condition $V|_\Gamma = 0$ we obtain the boundary conditions $X(0) = X(b_1) = 0$ for equation (3.12) and $Y(0) = Y(b_2) = 0$ for equation (3.13).

We have already encountered several times the Sturm-Liouville problem

$$\begin{cases} X'' + \mu X = 0, & 0 < x < b_1, \\ X(0) = X(b_1) = 0, \end{cases}$$

so we can immediately write

$$\mu_n = \left(\frac{n\pi}{b_1}\right)^2, \quad \text{and} \quad X_n(x) = \sin\left(\frac{n\pi}{b_1}x\right), \qquad n = 1, 2 \ldots$$

In much the same way, equation (3.13) comes with the boundary conditions

$$Y(0) = Y(b_2) = 0$$

and has the solutions

$$\lambda_{nm} - \mu_n = \left(\frac{m\pi}{b_2}\right), \quad Y_m(y) = \sin\left(\frac{m\pi}{b_2}y\right), \quad n = 1, 2, \ldots, \quad m = 1, 2, \ldots$$

Therefore,

$$\lambda_{nm} = \left(\frac{n\pi}{b_1}\right)^2 + \left(\frac{m\pi}{b_2}\right)^2$$

and equation (3.9) has the solution "atoms"

$$V_{nm} = \sin\left(\frac{n\pi}{b_1}x\right)\left(\frac{m\pi}{b_2}y\right).$$

Now from equation (3.10) is follows that

$$T_{nm} = c_{nm}e^{-\left[\left(\frac{n\pi a}{b_1}\right)^2 + \left(\frac{m\pi a}{b_2}\right)^2\right]t}.$$

We conclude that the solution of the original problem is written as a series

$$u(x, y, t) = \sum_{n=1}^{\infty}\sum_{m=1}^{\infty} c_{nm}e^{-\left[\left(\frac{n\pi a}{b_1}\right)^2 + \left(\frac{m\pi a}{b_2}\right)^2\right]t}\sin\left(\frac{n\pi}{b_1}x\right)\left(\frac{m\pi}{b_2}y\right).$$

The initial condition forces $c_{21} = 1$ and all the others $c_{nm} = 0$. Finally,

$$u(x, y, t) = e^{-\left[\frac{4\pi^2 a^2}{b_1^2} + \frac{\pi^2 a^2}{b_2^2}\right]t}\sin\left(\frac{2\pi}{b_1}x\right)\left(\frac{\pi}{b_2}y\right).$$

The case of a cylinder.

Example 5. Find the temperature of an bounded circular cylinder (Figure 3.5) whose surface is maintained at temperature zero and in which the initial temperature distributions is

$$u|_{t=0} = J_0\left(\frac{\mu_{05}}{b}\rho\right)\sin\left(\frac{2\pi}{h}z\right).$$

Solution. The mathematical formulation of the problem is as follows:

$$\begin{cases} u_t = a^2\Delta u, & 0 < \rho < b, \quad 0 < z < h, \quad t > 0, & (3.14) \\ u|_{t=0} = J_0\left(\frac{\mu_{05}}{b}\rho\right)\sin\left(\frac{2\pi}{h}z\right), & 0 \le \rho \le b, \quad 0 \le z \le h, & (3.15) \\ u|_{z=0} = u|_{z=h} = 0, & u|_{\rho=b} = 0, & t \ge 0. & (3.16) \end{cases}$$

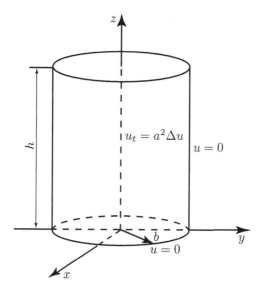

FIGURE 3.5. Heat propagation in a cylinder

Here

$$\Delta u = \frac{1}{\rho}\frac{\partial}{\partial\rho}\left(\rho\frac{\partial u}{\partial\rho}\right) + \frac{1}{\rho^2}\frac{\partial^2 u}{\partial\varphi^2} + \frac{\partial^2 u}{\partial z^2}.$$

Since the initial and boundary conditions are independent of φ, the solution of the problem will also be independent of φ, i.e., $u = u(\rho, z, t)$. Let us seek it in the form

$$u(\rho, z, t) = R(\rho)Z(z)T(t).$$

Substituting this expression in equation (3.14)

$$u_t = a^2\left[\frac{1}{\rho}\frac{\partial}{\partial\rho}\left(\rho\frac{\partial u}{\partial\rho}\right) + \frac{\partial^2 u}{\partial z^2}\right]$$

we obtain

$$RZT' = a^2\left[ZT\frac{1}{\rho}\frac{d}{d\rho}(\rho R') + RTZ''\right],$$

which upon dividing both side by RZT yields

$$\frac{T'}{a^2 T} = \frac{\frac{1}{\rho}\frac{d}{d\rho}(\rho R')}{R} + \frac{Z''}{Z} = -\lambda.$$

In this way we obtain two equations:

$$\frac{\frac{1}{\rho}\frac{d}{d\rho}(\rho R')}{R} + \frac{Z''}{Z} = -\lambda, \tag{3.17}$$

and

$$T' + a^2 \lambda T = 0. \tag{3.18}$$

Further, equation (3.17) yields

$$\frac{\frac{1}{\rho}\frac{d}{d\rho}(\rho R')}{R} + \lambda = -\frac{Z''}{Z} = \mu. \tag{3.19}$$

Thus, we have to solve the Sturm-Liouville problem

$$\begin{cases} Z'' + \mu Z = 0, & 0 < z < h, \\ Z(0) = Z(h) = 0. \end{cases}$$

Its solutions are

$$\mu_n = \left(\frac{n\pi}{h}\right)^2, \qquad Z_n(z) = \sin\left(\frac{n\pi}{h}z\right), \qquad n = 1, 2, \dots$$

Now from (3.19) it follows that

$$\frac{\frac{1}{\rho}\frac{d}{d\rho}(\rho R')}{R} = \mu - \lambda,$$

or

$$\rho^2 R'' + \rho R' + \rho^2 \nu R = 0, \tag{3.20}$$

where we denoted $\mu - \lambda = -\nu$.

With equation (3.20) we are already familiar: it has the solution $R(\rho) = J_0(\sqrt{\nu}\rho)$, where J_0 is the Bessel function of first kind and index zero. From the boundary condition $u|_{\rho=b} = 0$ is follows that $R(b) = 0$, i.e., $J_0(\sqrt{\nu}b) = 0$, which gives the sequence

$$\nu_m = \left(\frac{\mu_{0m}}{b}\right)^2, \qquad m = 1, 2, \dots$$

We see that the eigenvalues and eigenfunctions of our problem are

$$\lambda_{nm} = \left(\frac{n\pi}{h}\right)^2 + \left(\frac{\mu_{0m}}{b}\right)^2$$

and respectively

$$V_{nm} = J_0\left(\frac{\mu_{0m}}{b}\rho\right)\sin\left(\frac{n\pi}{h}z\right).$$

Next, equation (3.18) has the corresponding sequence of solutions

$$T_{nm} = c_{nm}e^{-a^2\lambda_{nm}t}.$$

Therefore, the general solution of our problem is

$$u(\rho, z, t) = \sum_{n=1}^{\infty}\sum_{m=1}^{\infty} c_{nm}e^{-a^2\lambda_{nm}t} J_0\left(\frac{\mu_{0m}}{b}\rho\right)\sin\left(\frac{n\pi}{h}z\right).$$

The initial condition (3.15) implies that $c_{25} = 1$ and $c_{nm} = 0$ if $n \neq 2, m \neq 5$.

We conclude that the temperature distribution in the cylinder is given by the formula

$$u(\rho, z, t) = e^{-a^2\pi^2[(\frac{4}{h})^2+(\frac{\mu_{05}}{b})^2]t} J_0\left(\frac{\mu_{05}}{b}\rho\right)\sin\left(\frac{2\pi}{h}z\right).$$

Example 6 [4, Ch. 4, no. 28]. Find the temperature of an infinite circular cylinder under the assumptions that the initial temperature is

$$u|_{t=0} = u_0\left(1 - \frac{\rho^2}{\rho_0^2}\right), \qquad u_0 = \text{const},$$

and the surface of the cylinder is maintained at temperature zero. Under conditions of a regular regime, find an approximate expression for the average of the temperature over the cross section of the cylinder.

Solution. Mathematically, the problem is formulated as follows:

$$\begin{cases} u_t = a^2\left(u_{\rho\rho} + \frac{1}{\rho}u_\rho\right), & 0 < \rho < \rho_0, \quad 0 < t < \infty, \\[2mm] u(\rho, 0) = u_0\left(1 - \frac{\rho^2}{\rho_0^2}\right), & 0 \leq \rho \leq \rho_0, \\[2mm] u(\rho_0, t) = 0, & 0 \leq t < \infty. \end{cases} \qquad (3.21)$$

First, let us consider the auxiliary problem

$$\begin{cases} u_t = a^2\left(u_{\rho\rho} + \frac{1}{\rho}u_\rho\right), & 0 < \rho < \rho_0, \quad 0 < t < \infty., \\[2mm] u(\rho_0, t) = 0, & 0 \leq t < \infty. \end{cases} \qquad (3.22)$$

Proceeding as we did before, we shall seek its solution in the form

$$u(\rho, t) = R(\rho)T(t). \qquad (3.23)$$

Substituting the expression (3.23) in equation (3.22), we obtain

$$RT' = a^2T\left(R'' + \frac{1}{\rho}R'\right),$$

whence, upon dividing both sides by RT,

$$\frac{T'}{a^2 T} = \frac{R'' + \frac{1}{\rho}R'}{R} = -\lambda,$$

where λ is the separation constant.

Again, we obtain two ordinary differential equations:

$$\rho^2 R'' + \rho R' + \lambda \rho^2 R = 0, \tag{3.24}$$

$$T' + a^2 \lambda T = 0. \tag{3.25}$$

The general solution of equation (3.24) has the form

$$R(\rho) = C_1 J_0(\sqrt{\lambda}\rho) + C_2 Y_0(\sqrt{\lambda}\rho).$$

Since we must have $|R(0)| < \infty$, $C_2 = 0$ (otherwise, $|R(\rho)| \to \infty$ as $\rho \to 0$ because $|Y_0(\sqrt{\lambda}\rho)| \to \infty$ as $\rho \to 0$). Therefore,

$$R(\rho) = C_1 J_0(\sqrt{\lambda}\rho).$$

The boundary condition $u(\rho_0, t) = 0$ yields $R(\rho_0) = 0$, i.e., $J_0(\sqrt{\lambda}\rho_0) = 0$. We obtain the sequence $\lambda_n = \left(\frac{\mu_n}{\rho_0}\right)^2$, where μ_n, $n = 1, 2, \ldots$ are the positive roots of the equation $J_0(x) = 0$. The corresponding solutions are $R_n(\rho) = J_0\left(\frac{\mu_n}{\rho_0}\rho\right)$.

Now from equation (3.25) with $\lambda = \lambda_n$ we find

$$T_n(t) = C e^{-\left(\frac{\mu_n}{\rho_0}a\right)^2 t},$$

where C is an arbitrary constant.

Therefore, the solution "atoms" of problem (3.22) are the functions

$$u_n(\rho, t) = J_0\left(\frac{\mu_n}{\rho_0}\rho\right) e^{-\left(\frac{\mu_n}{\rho_0}a\right)^2 t}$$

and the solution of the original problem (3.21) is given by a series

$$u(\rho, t) = \sum_{n=1}^{\infty} A_n J_0\left(\frac{\mu_n}{\rho_0}\rho\right) e^{-\left(\frac{\mu_n}{\rho_0}a\right)^2 t}, \tag{3.26}$$

where the coefficients A_n are subject to determination; namely, they are found from the initial condition

$$u(\rho, 0) = u_0 \left(1 - \frac{\rho^2}{\rho_0^2}\right).$$

Setting $t = 0$ in (3.26), we get

$$u(\rho, 0) = \sum_{n=1}^{\infty} A_n J_0\left(\frac{\mu_n}{\rho_0}\rho\right).$$

Therefore,

$$A_n = \frac{2}{\rho_0^2 J_1^2(\mu_n)} \int_0^{\rho_0} \rho u_0 \left(1 - \frac{\rho^2}{\rho_0^2}\right) J_0\left(\frac{\mu_n}{\rho_0}\rho\right) d\rho.$$

But we have already encountered such an integral, so we know the answer:

$$A_n = \frac{8u_0}{\mu_n^3 J_1(\mu_n)}.$$

We conclude that the solution of the original problem is

$$u(\rho, t) = 8u_0 \sum_{n=1}^{\infty} \frac{1}{\mu_n^3 J_1(\mu_n)} J_0\left(\frac{\mu_n}{\rho_0}\rho\right) e^{-\left(\frac{\mu_n}{\rho_0}a\right)^2 t}.$$

Let us note that since the roots of the Bessel function $J_0(x)$ increase with n, obeying the formula

$$\mu_n = \pi\left[n - \frac{1}{4} + \frac{0.56}{4n-1} + \frac{0.05}{(4n-1)^2} + \dots\right],$$

when $t \to \infty$ the leading term in the above sum formula for the solution is the first one, so that for sufficiently large values of t we have

$$u(\rho, t) \approx 8u_0 \frac{J_0\left(\frac{\mu_1}{\rho_0}\rho\right)}{\mu_n^3 J_1(\mu_1)} e^{-\left(\frac{\mu_1}{\rho_0}a\right)^2 t}.$$

The cross-section average of the temperature is

$$\bar{u}(t) \simeq \frac{1}{\pi\rho_0^2} \int_0^{\rho_0} u(\rho, t) 2\pi\rho \, d\rho \approx$$

$$\approx \frac{16u_0}{\mu_1^3 \rho_0^2 J_1(\mu_1)} e^{-\left(\frac{\mu_1}{\rho_0}a\right)^2 t} \int_0^{\rho_0} \rho J_0\left(\frac{\mu_1}{\rho_0}\rho\right) d\rho = \frac{16u_0}{\mu_1^4} e^{-\left(\frac{\mu_1}{\rho_0}a\right)^2 t},$$

because

$$\int_0^{\rho_0} \rho J_0\left(\frac{\mu_n}{\rho_0}\rho\right) d\rho = \frac{\rho_0^2}{\mu_1} J_1(\mu_1).$$

The case of a ball.

Example 7. Solve the following problem of heat propagation in a ball of radius b centered at the origin:

$$\begin{cases} u_t = \Delta u, & 0 < \rho < b, \quad 0 < \theta < \pi, \quad 0 < \varphi < 2\pi, \quad t > 0, \\[2mm] u|_{t=0} = \dfrac{3}{2}\sin(2\theta)\cos\varphi \dfrac{J_{5/2}\left(\frac{\mu_{2,1}}{b}\rho\right)}{\sqrt{\rho}}, \\[3mm] \qquad\qquad 0 \le \rho \le b, \quad 0 \le \theta \le \pi, \quad 0 \le \varphi \le 2\pi, \quad t > 0, \\[2mm] u|_\Gamma = 0, & \rho = b, \quad 0 \le \theta \le \pi, \quad 0 \le \varphi \le 2\pi, \quad t > 0, \end{cases}$$

where $\mu_{2,1}$ is the first positive root of the equation $J_{5/2}(x) = 0$.

Solution. Let us write the Laplace operator in spherical coordinates:

$$\Delta u = \frac{1}{\rho^2}\frac{\partial}{\partial\rho}\left(\rho^2\frac{\partial u}{\partial\rho}\right) + \frac{1}{\rho^2}\frac{1}{\sin^2\theta}\frac{\partial}{\partial\theta}\left(\sin\theta\frac{\partial u}{\partial\theta}\right) + \frac{1}{\rho^2\sin\theta}\frac{\partial u}{\partial\varphi^2}.$$

We will seek the solution of our problem in the form

$$u(\rho,\theta,\varphi,t) = T(t)R(\rho)Y(\theta,\varphi).$$

Substituting this expression in the heat equation we obtain

$$RYT' = a^2\frac{1}{\rho^2}\left[TY\frac{d}{d\rho}(\rho^2 R') + RT\frac{1}{\sin\theta}\frac{\partial}{\partial\theta}\left(\sin\theta\frac{\partial Y}{\partial\theta}\right) + RT\frac{1}{\sin^2\theta}\frac{\partial^2 Y}{\partial\varphi^2}\right].$$

This yields the equality

$$\frac{T'}{a^2 T} = \frac{1}{\rho^2}\left[\frac{\frac{d}{d\rho}(\rho^2 R')}{R} + \frac{\frac{1}{\sin\theta}\frac{\partial}{\partial\theta}\left(\sin\theta\frac{\partial Y}{\partial\theta}\right) + RT\frac{1}{\sin^2\theta}\frac{\partial^2 Y}{\partial\varphi^2}}{Y}\right] = -\lambda,$$

which in turn results in the following two equations:

$$\frac{1}{\rho^2}\left[\frac{\frac{d}{d\rho}(\rho^2 R')}{R} + \frac{\Delta_{\theta\varphi}Y}{Y}\right] = -\lambda, \tag{3.27}$$

and

$$T' + a^2\lambda T = 0. \tag{3.28}$$

Here we denoted by $\Delta_{\theta\varphi}$ the operator

$$\Delta_{\theta\varphi}Y = \frac{1}{\sin\theta}\frac{\partial}{\partial\theta}\left(\sin\theta\frac{\partial Y}{\partial\theta}\right) + \frac{1}{\sin^2\theta}\frac{\partial^2 Y}{\partial\varphi^2}.$$

From equation (3.27) it follows that

$$\frac{\frac{d}{d\rho}(\rho^2 R')}{R} + \lambda\rho^2 = -\frac{\Delta_{\theta\varphi}Y}{Y} = \mu. \tag{3.29}$$

Here λ and μ are separation constants. Next, (3.29) gives the equation

$$\Delta_{\theta\varphi}Y + \mu Y = 0. \tag{3.30}$$

to which we must add the "boundary" conditions

$$\begin{cases} Y(\theta,\varphi) = Y(\theta,\varphi+2\pi), \\ |Y(0,\varphi)| < \infty, \quad |Y(\pi,\varphi)| < \infty. \end{cases} \tag{3.31}$$

In this way we obtain the eigenvalue problem (3.30),(3.31): find those values of the parameter μ for which the problem has nontrivial solutions, and then find those solutions. We shall seek the solutions in the form

$$Y(\theta,\varphi) = A(\theta)\Phi(\varphi).$$

First of all, notice that from the equality $Y(\theta,\varphi) = Y(\theta,\varphi+2\pi)$ it follows that

$$\Phi(\varphi) = \Phi(\varphi+2\pi). \tag{3.32}$$

Then equation (3.30) reads

$$\Phi\frac{1}{\sin\theta}\frac{\partial}{\partial\theta}(\sin\theta \cdot A'(\theta)) + \frac{1}{\sin^2\theta}\Phi''A + \mu A\Phi = 0,$$

whence

$$\frac{\frac{1}{\sin\theta}\frac{\partial}{\partial\theta}(\sin\theta \cdot A')}{A} + \frac{1}{\sin^2\theta}\frac{\Phi''}{\Phi} = -\mu,$$

or

$$\frac{\sin\theta\frac{d}{d\theta}(\sin\theta \cdot A')}{A} + \mu\sin^2\theta + \frac{\Phi''}{\Phi} = 0,$$

or

$$\frac{\sin\theta\frac{d}{d\theta}(\sin\theta \cdot A')}{A} + \mu\sin^2\theta = -\frac{\Phi''}{\Phi} = \nu,$$

where ν is a new separation constant. This yields the following problem for Φ:

$$\Phi'' + \nu\Phi = 0, \qquad -\infty < \varphi < \infty, \tag{3.33}$$

$$\Phi(\varphi) = \Phi(\varphi+2\pi). \tag{3.34}$$

It is readily verified that its eigenvalues are

$$\nu = m^2, \qquad m = 0, 1, 2, \ldots$$

and the corresponding solutions are

$$\Phi_m(\varphi) = \begin{cases} \cos(m\varphi), \\ \sin(m\varphi). \end{cases}$$

We also have the following problem from which one determines $A(\theta)$:

$$\begin{cases} \sin\theta \dfrac{d}{d\theta}(\sin\theta \cdot A') + (\mu \sin^2\theta - m^2)A = 0, \qquad 0 < \theta < \pi, \\ |A(0)| < \infty, \quad |A(\pi)| < \infty. \end{cases}$$

Let us denote $x = \cos\theta$. The problem for A reduces to the form

$$\frac{d}{dx}\left[(1 - x^2)\frac{dy}{dx}\right] + \left(\mu - \frac{m^2}{1 - x^2}\right)y = 0 \qquad -1 < x < 1, \tag{3.35}$$

$$|y(1)| < \infty, \quad |y(-1)| < \infty, \tag{3.36}$$

where $y(x)|_{x=\cos\theta} = A(\theta)$.

We recognize this to be a boundary value problem for the associated Legendre equation. It is known that the eigenvalues of problem (3.35), (3.36) are $\mu_n = n(n+1)$, where $n = 0, 1, 2, \ldots$, and the corresponding eigenfunctions are given by

$$P_n^{(m)}(x) = (1 - x^2)^{m/2}\frac{d^m P_n(x)}{dx^m},$$

where $P_n(x)$ is the Legendre polynomial of degree n (here $m \leq n$); $P_n^{(m)}$ are called the *associated Legendre polynomials* of order m.

We conclude that the solutions of equation (3.30) under the conditions (3.31) are the *spherical functions*

$$P_n^{(0)}(\cos\theta), \quad P_n^{(m)}(\cos\theta)\cos(m\varphi),$$
$$P_n^{(m)}(\cos\theta)\sin(m\varphi), \qquad m = 1, 2, \ldots, n.$$

Now let us return to relation (3.29). Taking $\mu = n(n+1)$ and imposing the boundary condition $R(b) = 0$, we obtain the following boundary value problem for the function $R(\rho)$:

$$\begin{cases} \rho^2 R'' + 2\rho R' + [\lambda\rho^2 - n(n+1)]R = 0, \\ |R(0)| < \infty, \quad R(b) = 0. \end{cases} \tag{3.37}$$

The substitution $R(\rho) = y(\rho)/\sqrt{\rho}$ transforms (3.37) into the Bessel equation of half-integer index

$$\rho^2 y'' + \rho y' + \left[\lambda \rho^2 - \left(n + \frac{1}{2}\right)^2\right] y = 0,$$

whose general solution has the form

$$y(\rho) = C_1 J_{n+1/2}(\sqrt{\lambda}\rho) + C_1 N_{n+1/2}(\sqrt{\lambda}\rho).$$

Here $J_{n+1/2}(x)$ and $N_{n+1/2}(x)$ are the Bessel functions of the first and second kind, respectively, of index $n + 1/2$, and C_1, C_2 are arbitrary constants. Clearly, for our problem we must put $C_2 = 0$, i.e.,

$$y(\rho) = C_1 J_{n+1/2}(\sqrt{\lambda}\rho).$$

Imposing the condition $R(b) = 0$ we obtain

$$J_{n+1/2}(\sqrt{\lambda}b) = 0.$$

Therefore,

$$\lambda_{nk} = \left(\frac{\mu_{nk}}{b}\right)^2,$$

where μ_{nk}, $k = 1, 2, \dots$, are the positive roots of the equation

$$J_{n+1/2}(x) = 0.$$

Is it obvious that to each eigenvalue λ_{nk} there correspond $2n + 1$ eigenfunctions

$$v_{nkj} = \frac{J_{n+1/2}\left(\frac{\mu_{nk}}{b}\rho\right)}{\sqrt{\rho}} Y_n^{(j)}(\theta, \varphi), \qquad j = 0, \pm 1, \pm 2, \dots, \pm n,$$

where

$$Y_n^{(m)}(\theta, \varphi) = P_n^{(m)}(\cos\theta)\cos(m\varphi),$$
$$Y_n^{(-m)}(\theta, \varphi) = P_n^{(m)}(\cos\theta)\sin(m\varphi),$$
$$m = 0, 1, 2, \dots, n.$$

Further, from equation (2) it follows that

$$T_{nk} = C_{nk} e^{-\left(\frac{a\mu_{nk}}{b}\right)^2 t},$$

where C_{nk} is a constant. This yield solution "atoms" of our problem:

$$u_{nk}(\rho, \theta, \varphi, t) = \frac{J_{n+1/2}\left(\frac{\mu_{nk}}{b}\rho\right)}{\sqrt{\rho}} Y_n(\theta, \varphi) e^{-\left(\frac{a\mu_{nk}}{b}\right)^2 t}.$$

Here the spherical harmonic $Y_n(\theta, \varphi)$ has the expression

$$Y_n(\theta, \varphi) = \sum_{m=0}^{n} [A_{nm} \cos(m\varphi) + B_{nm} \sin(m\varphi)] P_n^{(m)}(\cos \theta),$$

with constants A_{nm} and B_{nm}. Thus, the solution of our problem is

$$u(\rho, \theta, \varphi, t) = \sum_{n=0}^{\infty} \sum_{k=1}^{\infty} \frac{J_{n+1/2}\left(\frac{\mu_{nk}}{b}\rho\right)}{\sqrt{\rho}} Y_n(\theta, \varphi) e^{-\left(\frac{a\mu_{nk}}{b}\right)^2 t}.$$

Imposing the initial condition, we finally have

$$u(\rho, \theta, \varphi, t) = \frac{J_{5/2}\left(\frac{\mu_{21}}{b}\rho\right)}{\sqrt{\rho}} P_2^{(1)}(\cos \theta) e^{-\left(\frac{a\mu_{21}}{b}\right)^2 t}.$$

3.4. A modification of the method of separation of variables for solving the Cauchy problem

The solution of the Cauchy problem

$$\begin{cases} u_t = a^2 \Delta u + f(x, y, z, t), & -\infty < x, y, z < \infty, \quad t > 0, \\ u|_{t=0} = g(x, y, z), & -\infty < x, y, z < \infty, \end{cases}$$

is provided (generally speaking) by the Poisson integral. But often the calculations connected with the Poisson integral are so tedious that in practice one decides not to use it. As it turns out, for a wide class of functions f one can apply the method of particular solutions. To this end let us observe that the heat operator $Lu = u_t - \Delta u$ takes (i.e., maps), for example, a function of the form $\varphi(t) \sin x \cos y$ (in the two-dimensional case) into a function of the same form $\psi(t) \sin x \cos y$.

Let us examine a number of examples.

Example 1 [6, Ch. IV, 13.5(3)]. Solve the Cauchy problem

$$\begin{cases} u_t = u_{xx} + e^{-t} \cos x, & -\infty < x < \infty, \quad t > 0, \\ u|_{t=0} = \cos x, & -\infty < x < \infty. \end{cases}$$

Solution. As we have indicated, the solution may be obtained by Poisson's formula, but sometimes (as in the present case) is it more convenient to apply the *method of separation of variables*. To that end we split our problem into two simpler problems:

$$(1) \quad \begin{cases} v_t = v_{xx} + e^{-t} \cos x, & -\infty < x < \infty, \quad t > 0, \\ v|_{t=0} = 0, & -\infty < x < \infty \end{cases}$$

and

$$(2) \quad \begin{cases} w_t = w_{xx}, & -\infty < x < \infty, \quad t > 0, \\ w|_{t=0} = \cos x, & -\infty < x < \infty \end{cases}$$

Let us seek the function $v(x,t)$ in the form

$$v(x,t) = \varphi(t) \cos x,$$

where the function $\varphi(t)$ needs to be determined. Substituting this expression in the equation of problem (1), we obtain

$$\varphi'(t) \cos x = -\varphi(t) \cos x + e^{-t} \cos x,$$

which yields the following Cauchy problem for an ordinary differential equation (where we use the fact that $\varphi(0) = 0$):

$$\begin{cases} \varphi' + \varphi = e^{-t}, & t > 0, \\ \varphi(0) = 0. \end{cases}$$

It is readily verified that the solution of this problem is

$$\varphi(t) = te^{-t}.$$

Therefore, $v(x,t) = te^{-t} \cos x$. Further, the solution of problem (2) is the function

$$w(x,t) = e^{-t} \cos x.$$

We conclude that

$$u(x,t) = v(x,t) + w(x,t) = e^{-t}(1+t) \cos x.$$

Example 2 [6, Ch. IV, 13.6(2)]. Solve the Cauchy problem

$$\begin{cases} u_t = \Delta u + \sin t \sin x \sin y, & -\infty < x, y < \infty, \quad t > 0, \\ u|_{t=0} = 1, & -\infty < x, y < \infty. \end{cases}$$

Solution. As before, let us split the problem into two problems:

$$(1) \quad \begin{cases} v_t = \Delta v + \sin t \sin x \sin y, & -\infty < x, y < \infty, \quad t > 0, \\ v|_{t=0} = 0, & -\infty < x, y < \infty, \end{cases}$$

and

$$(2) \quad \begin{cases} w_t = \Delta w, & -\infty < x, y < \infty, \quad t > 0, \\ w|_{t=0} = 1, & -\infty < x, y < \infty. \end{cases}$$

We see right away that $w(x,y,t) = 1$.

As for the function $v(x, y, t)$, let us seek it in the form

$$v(x, y, t) = \varphi(t) \sin x \sin y.$$

Substituting this expression in the equation of problem (1) we obtain

$$\varphi'(t) \sin x \sin y = -2\varphi(t) \sin x \sin y + \sin t \sin x \sin y,$$

which in turn gives the following Cauchy problem for the function $\varphi(t)$:

$$\begin{cases} \varphi' + 2\varphi = \sin t, & t \geq 0, \\ \varphi(0) = 0. \end{cases}$$

Its solution is the function

$$\varphi(t) = \frac{1}{5}(2 \sin t - \cos t + e^{-2t}),$$

and so

$$v(x, y, t) = \frac{1}{5} \sin x \sin y (2 \sin t - \cos t + e^{-2t}).$$

We conclude that

$$u(x, y, t) = v(x, y, t) + w(x, y, t) = \frac{1}{5} \sin x \sin y (2 \sin t - \cos t + e^{-2t}) + 1.$$

It also turns out that if the initial data of the problem have some specific symmetry properties, then it is natural to search for the solution of the Cauchy problem in the same class of functions, i.e., among the functions with the same symmetry properties. Let us give some relevant examples.

Example 3 [6, Ch. IV, 13.8(5)]. Solve the Cauchy problem

$$\begin{cases} u_t = \Delta u, & x = (x_1, \dots, x_n) \in \mathbf{R}^n, \quad t > 0, \\ u|_{t=0} = e^{-(\sum_{k=1}^n x_k)^2}, & x \in \mathbf{R}^n \end{cases}$$

(here $\mathbf{R}^n$ is the n-dimensional Euclidean space).

Solution. We shall assume that

$$u(x_1, \dots, x_n, t) = v(x_1 + \dots + x_n; t).$$

Then denoting $s = x_1 + \dots + x_n$ and observing that $u_t = v_t$, we have that

$$\frac{\partial^2 u}{\partial x_i^2} = \frac{\partial^2 v}{\partial s^2},$$

whence

$$\Delta u = \frac{\partial^2 u}{\partial x_1^2} + \ldots + \frac{\partial^2 u}{\partial x_n^2} = n\frac{\partial^2 v}{\partial s^2}.$$

Therefore, the equation $u_t = \Delta u$ is recast as

$$v_t = n\frac{\partial^2 v}{\partial s^2}$$

and the original Cauchy problem becomes

$$\begin{cases} v_t = n\dfrac{\partial^2 v}{\partial s^2}, & -\infty < s < \infty, \quad t > 0, \\ v|_{t=0} = e^{-s^2}, & -\infty < s < \infty. \end{cases}$$

Applying here the Poisson integral we have

$$v(s,t) = \frac{1}{2\sqrt{n\pi t}} \int_{-\infty}^{\infty} e^{-\zeta^2} e^{-\frac{(s-\zeta)^2}{4nt}} \, d\zeta =$$

$$= \frac{1}{2\sqrt{n\pi t}} e^{-\frac{s^2}{4nt}} \int_{-\infty}^{\infty} e^{-\zeta^2 + \frac{s\zeta}{2nt} - \frac{\zeta^2}{4nt}} \, d\zeta =$$

$$= \frac{1}{2\sqrt{n\pi t}} e^{-\frac{s^2}{1+4nt}} \int_{-\infty}^{\infty} e^{-\frac{1+4nt}{4nt}\left(\zeta - \frac{s}{1+4nt}\right)^2} \, d\zeta.$$

To compute the last integral we make the change of variables

$$\alpha = \sqrt{\frac{1+4nt}{4nt}} \left(\zeta - \frac{s}{1+4nt}\right).$$

Then we have

$$v(s,t) = \frac{1}{2\sqrt{n\pi t}} \sqrt{\frac{1+4nt}{4nt}} e^{-\frac{s^2}{1+4nt}} \int_{-\infty}^{\infty} e^{-\alpha^2 d\alpha} \, d\alpha = \frac{1}{\sqrt{1+4nt}} e^{-\frac{s^2}{1+4nt}}.$$

Therefore, the solution of the original Cauchy problem is given by the formula

$$u(x_1, \ldots, x_n, t) = \frac{1}{\sqrt{1+4nt}} e^{-\frac{1}{1+4nt}\left(\sum_{k=1}^{n} x_k\right)^2}.$$

Example 4. [6, Ch. IV, 13.6(4)]. Solve the Cauchy problem

$$\begin{cases} 8u_t = \Delta u + 1, & -\infty < x, y < \infty, \quad t > 0, \\ u|_{t=0} = e^{-(x-y)^2}, & -\infty < x, y < \infty. \end{cases}$$

Solution. Let us seek the solution as the sum of the solutions of two problems:

$$(1) \quad \begin{cases} 8u_t^{(1)} = \Delta u^{(1)} + 1, & -\infty < x, y < \infty, \quad t > 0, \\ u^{(1)}|_{t=0} = 0, & -\infty < x, y < \infty, \end{cases}$$

and

$$(2) \quad \begin{cases} 8u_t^{(2)} = \Delta u^{(2)}, & -\infty < x, y < \infty, \quad t > 0, \\ u^{(2)}|_{t=0} = e^{-(x-y)^2}, & -\infty < x, y < \infty. \end{cases}$$

The solution of problem (1) has the form $u^{(1)}(x, y, t) = \frac{1}{8}t$. Let us find the solution of problem (2). We shall assume that

$$u^{(2)}(x, y, t) = v(x - y; t).$$

Then denoting $x - y = z$ and observing that

$$\frac{\partial u^{(2)}}{\partial x} = \frac{\partial v}{\partial z}\frac{\partial z}{\partial x} = \frac{\partial v}{\partial z},$$

$$\frac{\partial^2 u^{(2)}}{\partial x^2} = \frac{\partial^2 v}{\partial z^2}, \qquad \frac{\partial^2 u^{(2)}}{\partial y^2} = \frac{\partial^2 v}{\partial z^2},$$

we obtain the Cauchy problem

$$\begin{cases} v_t = \dfrac{1}{4}\dfrac{\partial^2 v}{\partial z^2}, & -\infty < z < \infty, \quad t > 0, \\ v|_{t=0} = e^{-z^2}, & -\infty < z < \infty. \end{cases}$$

The solution of this problem is given by the Poisson formula (with $a = 1/2$):

$$v(z, t) = \frac{1}{\sqrt{\pi t}} \int_{-\infty}^{\infty} e^{-\zeta^2} e^{-\frac{(z-\zeta)^2}{t}} d\zeta = \frac{1}{\sqrt{\pi t}} e^{-\frac{z^2}{t}} \int_{-\infty}^{\infty} e^{-\zeta^2 + \frac{2z\zeta}{t} - \frac{\zeta^2}{t}} d\zeta =$$

$$= \frac{1}{\sqrt{\pi t}} e^{-\frac{z^2}{t}} \int_{-\infty}^{\infty} e^{-\left[\left(1+\frac{1}{t}\right)\left(\zeta^2 - \frac{2z\zeta}{1+t} + \left(\frac{z}{1+t}\right)^2\right)\right]} e^{\frac{z^2}{t(1+t)}} d\zeta =$$

$$= \frac{1}{\sqrt{\pi t}} e^{-\frac{z^2}{1+t}} \sqrt{\frac{t}{1+t}} \int_{-\infty}^{\infty} e^{-\left(\frac{1+t}{t}\right)\left(\zeta - \frac{z}{1+t}\right)^2} \sqrt{\frac{t+1}{t}} d\zeta =$$

$$= \frac{1}{\sqrt{\pi}} e^{-\frac{z^2}{1+t}} \frac{1}{\sqrt{1+t}} \int_{-\infty}^{\infty} e^{-\alpha^2} d\alpha = e^{-\frac{z^2}{1+t}} \frac{1}{\sqrt{1+t}},$$

where we used the notation

$$\alpha = \sqrt{\frac{1+t}{t}}\left(\zeta - \frac{z}{t+1}\right).$$

Therefore,

$$u^{(2)}(x, y, t) = \frac{1}{\sqrt{1+t}} e^{-\frac{(x-y)^2}{1+t}},$$

and the solution of the original problem is

$$u(x, y, t) = \frac{1}{8}t + \frac{1}{\sqrt{1+t}} e^{-\frac{(x-y)^2}{1+t}}.$$

Example 5 [6, Ch. IV, 13.7(4)]. Solve the Cauchy problem

$$\begin{cases} u_t = \Delta u + \cos(x - y + z), & -\infty < x, y, z < \infty, \quad t > 0, \\ u|_{t=0} = e^{-(x+y-z)^2}, & -\infty < x, y, z < \infty. \end{cases}$$

Solution. Again, let us split the problem into two auxiliary problems:

$$(1) \quad \begin{cases} u_t^{(1)} = \Delta u^{(1)} + \cos(x - y + z), & -\infty < x, y, z < \infty, \quad t > 0, \\ u^{(1)}|_{t=0} = 0, & -\infty < x, y, z < \infty, \end{cases}$$

and

$$(2) \quad \begin{cases} u_t^{(2)} = \Delta u^{(2)}, & -\infty < x, y, z < \infty, \quad t > 0, \\ u^{(2)}|_{t=0} = e^{-(x+y-z)^2}, & -\infty < x, y, z < \infty. \end{cases}$$

Let us seek solution of problem (1) in the form

$$u^{(1)}(x, y, z, t) = \varphi(t) \cos(x - y + z),$$

where the function $\varphi(t)$ needs to be determined. Substituting this expression for $u^{(1)}(x, y, z, t)$ in the equation of problem (1) we obtain

$$\varphi'(t) \cos(x - y + z) = 3\varphi(t)[-\cos(x - y + z)] + \cos(x - y + z),$$

because $\Delta \cos(x - y + z) = (-3) \cos(x - y + z)$. It follows that $\varphi(t)$ is the solution of the Cauchy problem

$$\begin{cases} \varphi'(t) + 3\varphi(t) = 1, & t > 0, \\ \varphi(0) = 0. \end{cases}$$

Clearly, the solution of this problem is $\varphi(t) = \frac{1}{3}(1 - e^{-3t})$, and so

$$u^{(1)}(x, y, z, t) = \frac{1}{3}(1 - e^{-3t}) \cos(x - y + z).$$

Further, to find the solution of problem (2) we use the representation

$$u^{(2)}(x, y, z, t) = v(x + y - z; t).$$

Then, observing that $\Delta u^{(2)} = \Delta v = 3\partial^2 v/\partial s^2$, where $s = x + y - z$, we recast problem (2) as

$$\begin{cases} v_t = 3\dfrac{\partial^2 v}{\partial s^2}, & -\infty < s < \infty, \quad t > 0, \\ v|_{t=0} = e^{-s^2}, & -\infty < s < \infty. \end{cases}$$

Poisson's formula (with $a = \sqrt{3}$) yields

$$v(s,t) = \frac{1}{2\sqrt{3}\sqrt{\pi t}} \int_{-\infty}^{\infty} e^{-\zeta^2} e^{-\frac{(s-\zeta)^2}{12t}} \, d\zeta =$$

$$= \frac{1}{2\sqrt{3}\sqrt{\pi t}} e^{-\frac{s^2}{12t}} \int_{-\infty}^{\infty} e^{-\zeta^2 + \frac{s\zeta}{6t} - \frac{\zeta^2}{12t}} \, d\zeta =$$

$$= \frac{1}{2\sqrt{3}\sqrt{\pi t}} e^{-\frac{s^2}{12t}} \int_{-\infty}^{\infty} e^{-\left[\left(1+\frac{1}{12t}\right)\left(\zeta^2 - \frac{2s\zeta}{1+12t} + \frac{s^2}{(1+12t)^2}\right)\right]} e^{\frac{s^2}{12t(1+12t)}} \, d\zeta =$$

$$= \frac{1}{2\sqrt{3\pi t}} e^{-\frac{s^2}{1+12t}} \int_{-\infty}^{\infty} e^{-\left(\frac{1+12t}{12t}\right)\left(\zeta - \frac{s}{1+12t}\right)^2} \, d\zeta =$$

$$= \frac{1}{2\sqrt{3\pi t}} e^{-\frac{s^2}{1+12t}} \sqrt{\frac{12t}{1+12t}} \int_{-\infty}^{\infty} e^{-\left(\frac{1+12t}{12t}\right)\left(\zeta - \frac{s}{1+12t}\right)^2} \sqrt{\frac{1+12t}{12t}} \, d\zeta =$$

$$\frac{1}{\sqrt{\pi}} \frac{1}{\sqrt{1+12t}} e^{-\frac{s^2}{1+12t}} \int_{-\infty}^{\infty} e^{-\alpha^2} \, d\alpha = \frac{1}{\sqrt{1+12t}} e^{-\frac{s^2}{1+12t}},$$

where we used the notation

$$\alpha = \sqrt{\frac{1+12t}{12t}} \left(\zeta - \frac{s}{1+12t}\right).$$

Therefore,

$$u^{(2)}(x,y,z,t) = \frac{1}{\sqrt{1+12t}} e^{-\frac{(x+y-z)^2}{1+12t}}$$

and the solution our problem is

$$u(x,y,z,t) = \frac{1}{3}(1 - e^{-3t})\cos(x - y + z) + \frac{1}{\sqrt{1+12t}} e^{-\frac{(x+y-z)^2}{1+12t}}.$$

3.5. Problems for independent study

1. Solve the Cauchy problem

$$\begin{cases} u_t = a^2 u_{xx}, & -\infty < x < \infty, \quad t > 0, \\ u(x,0) = f(x), & -\infty < x < \infty. \end{cases}$$

2. Solve the boundary value problem

$$\begin{cases} u_t = a^2 u_{xx}, & 0 < x, t < \infty, \\ u(0,t) = 0, & 0 \le t < \infty, \\ u(x,0) = f(x), & 0 \le x < \infty. \end{cases}$$

3. Solve the boundary value problem

$$\begin{cases} u_t = a^2 \Delta u, & -\infty < x, y < \infty, \quad 0 < z < \infty, \quad t > 0, \\ u|_{z=0} = 0, & -\infty < x, y < \infty, \quad t > 0, \\ u|_{t=0} = f(x,y,z), & -\infty < x, y < \infty, \quad 0 < z < \infty. \end{cases}$$

4. Using the result of problem **1**, show (applying Duhamel's method) that the solution of the Cauchy problem

$$\begin{cases} u_t = a^2 u_{xx} + f(x,t), & -\infty < x < \infty, \quad t > 0, \\ u|_{t=0} = 0, & -\infty < x < \infty, \end{cases}$$

is given by the function

$$u(x,t) = \frac{1}{2a\sqrt{\pi}} \int_0^t \frac{d\tau}{\sqrt{t-\tau}} \int_{-\infty}^{\infty} f(\zeta,\tau) e^{-\frac{(z-\zeta)^2}{4a^2(t-\tau)}} \, d\zeta.$$

5. Solve the boundary value problem

$$\begin{cases} u_t = u_{xx} + a^2 u + f(x), & 0 < x < \infty, \quad 0 < t < \infty, \\ u(0,t) = 0, \quad u_x(0,t) = 0, & 0 \le t < \infty. \end{cases}$$

6. Solve the boundary value problem

$$\begin{cases} u_t = u_{xx} + u + a \cos x, & 0 < x < \infty, \quad 0 < t < \infty, \\ u(0,t) = be^{-3t}, \quad u_x(0,t) = 0, & 0 \le t < \infty. \end{cases}$$

7. The initial temperature of a thin homogeneous rod is equal to zero. Determine the temperature $u(x,t)$ in the rod for $t > 0$ in the case when the rod is semi-infinite $(0 < x < \infty)$ and $u(0,t) = \delta(t)$.

Hint. The Laplace transform of the δ-function is 1.

8. The initial temperature of a thin homogeneous rod of length l is equal to $u(x,0) = 2\sin(3x) + 5\sin(8x)$. Determine the temperature $u(x,t)$ in the rod for $t > 0$ if the extremities are kept at the temperature of melting ice.

9. The initial temperature of a thin homogeneous rod of length $l = \pi$ is equal to $u(x,0) = 2\cos x + 3\cos(2x)$. Determine the temperature $u(x,t)$ in the rod for $t > 0$ if the heat flux at the extremities of the rod is equal to zero.

10. Solve the boundary value problem

$$\begin{cases} u_t = a^2 u_{xx} + u + t\sin(2x), & 0 < x < \pi, \quad t > 0, \\ u(x,0) = 0, & 0 \le x \le \pi, \\ u(0,t) = u(\pi,t) = 0, & t \ge 0. \end{cases}$$

11. Prove the uniqueness of the solution of the mixed problem

$$\begin{cases} u_t = a^2 u_{xx} + u + f(x,t), & 0 < x < l, \quad t > 0, \\ u(x,0) = \alpha(t), \quad u(l,t) = \beta(t), & t \ge 0, \\ u(0,t) = g(x), & 0 \le x \le l. \end{cases}$$

12. Find the temperature distribution in a thin rectangular plate of dimensions $[0,\pi] \times [0,\pi]$ if the initial temperature is equal to $u(x,y,0) = 3\sin x \sin(5y)$ and its boundary is maintained at temperature zero.

13. Find the temperature distribution in a thin rectangular plate of dimensions $[0,\pi] \times [0,\pi]$ if the initial temperature is equal to $u(x,y,0) = 5\cos x \cos(3y)$ and the heat flux through its boundary is equal to zero.

14. Find the temperature distribution in a thin rectangular plate of dimensions $[0,\pi] \times [0,\pi]$ under null initial and boundary conditions, when inside the plate there are heat sources with density $f(x,y,t) = A\sin x \sin y$, where $A = \mathrm{const}$.

15. Solve the following mixed problem in the rectangle $[0,\pi] \times [0,\pi]$:

$$\begin{cases} u_t = a^2(u_{xx} + u_{yy}), & 0 < x,y < \pi, \quad t > 0, \\ u(0,y,t) = 0, \quad u(\pi,y,t) = \sin y, & t \ge 0, \\ u(x,0,t) = u(x,\pi,t) = 0, & t \ge 0, \\ u(x,y,0) = 0, & 0 \le x,y \le \pi. \end{cases}$$

16. Let the function $u_k(x,t)$ be the solution of the Cauchy problem $(k=1,2,\dots,n)$

$$\begin{cases} \dfrac{\partial u_k}{\partial t} = a^2 \dfrac{\partial^2 u_k}{\partial x_k^2}, & -\infty < x_k < \infty, \quad t > 0, \\ u_k|_{t=0} = f_k(x_k), & -\infty < x_k < \infty. \end{cases}$$

Show that the function $u(x,t) = \prod_{k=1}^{n} u_k(x_k, t)$ (where $\prod$ denotes the product operation) is the solution of the Cauchy problem

$$
\begin{cases}
\dfrac{\partial u_k}{\partial t} = a^2 \Delta u, & x \in \mathbf{R}^n, \quad t > 0, \\[2mm]
u|_{t=0} = \displaystyle\prod_{k=1}^{n} f_k(x_k), & x \in \mathbf{R}^n,
\end{cases}
$$

where $x = (x_1, \ldots, x_n)$, $u(x) = u(x_1, \ldots, x_n)$.

17. Suppose that the function $f(x,t)$ is harmonic in the variable x for each fixed $t \geq 0$. Show that the function

$$
u(x,t) = \int_0^t f(x, \tau)\, d\tau
$$

is the solution of the Cauchy problem

$$
\begin{cases}
u_t = a^2 \Delta u + f(x,t), & x \in \mathbf{R}^n, \quad t > 0, \\
u|_{t=0} = 0, & x \in \mathbf{R}^n.
\end{cases}
$$

18. Find the temperature distribution in a rod $0 \leq x \leq l$ with thermally insulated lateral surface if its right endpoint $x = l$ is maintained at temperature zero and the temperature in the left point $x = 0$ is equal to $u(0,t) = At$, where $A = \mathrm{const}$. The initial temperature of the rod is assumed to be equal to zero.

19. Find the temperature distribution inside an infinite $(-\infty < z < \infty)$ circular cylinder of radius R if its surface is maintained at temperature zero and the initial temperature distribution is $u|_{t=0} = A J_0 \left(\frac{\mu_k}{R} \rho \right)$, where μ_k is the kth positive root of the equation $J_0(\mu) = 0$ and $A = \mathrm{const}$.

20. Find the temperature distribution inside an infinite $(-\infty < z < \infty)$ circular cylinder of radius R if its surface is maintained at temperature u_0 and the initial temperature is $u|_{t=0} = 0$.

21. Find the temperature distribution inside an infinite $(-\infty < z < \infty)$ circular cylinder of radius a if its surface is maintained at temperature zero and the initial temperature distribution is $u|_{t=0} = A J_2 \left(\frac{\mu_3}{a} \rho \right) \cos(2\varphi)$, where μ_3 is the third positive root of the Bessel function $J_2(x)$ and $A = \mathrm{const}$.

22. Find the temperature distribution in an infinite $(-\infty < z < \infty)$ circular cylinder of radius a if its surface is maintained at temperature zero, the initial temperature is equal to zero, and heat sources with density $f(\rho, t) = A J_0 \left(\frac{\mu_1}{a} \rho \right)$, act inside the cylinder, where μ_1 is the first positive root of the Bessel function $J_0(x)$ and $A = \mathrm{const}$.

23. Solve the mixed problem for the heat equation in a bounded cylinder

$$
\begin{cases}
u_t = c^2 \Delta u + e^t J_0 \left(\dfrac{\mu_1}{a} \rho \right) \cos \left(3 \dfrac{\pi}{h} z \right), & 0 < \rho < a, \quad 0 < x < h, \quad t > 0, \\
\dfrac{\partial u}{\partial n}\bigg|_\Gamma = 0, & t \geq 0, \\
u|_{t=0} = 0, & 0 \leq \rho \leq a, \quad 0 \leq z \leq h.
\end{cases}
$$

Here Γ is the boundary of the cylinder and μ_1 is the first positive root of the equation $J_0'(x) = 0$.

24. Determine the temperature inside an infinite $(-\infty < z < \infty)$ circular cylinder of radius a if the temperature of the lateral surface is $A \cos(2\varphi)$ $(A = \mathrm{const})$ and the initial temperature is equal to zero.

25. Determine the temperature inside an infinite $(-\infty < z < \infty)$ circular cylinder of radius a if the temperature of the lateral surface is $Ae^{-\omega^2 t}$ $(A = \mathrm{const})$ and the initial temperature is equal to zero.

26. Determine the temperature inside an infinite $(-\infty < z < \infty)$ circular cylinder of radius a if the temperature of the lateral surface is $Ae^{\omega^2 t}$ $(A = \mathrm{const})$ and the initial temperature is equal to zero.

27. Determine the temperature inside an infinite $(-\infty < z < \infty)$ circular cylinder of radius a if the temperature of its lateral surface is $Ae^{\omega^2 t} \sin(n\varphi)$ $(A = \mathrm{const})$ and the initial temperature is equal to zero; here n is an arbitrary positive integer.

28. Determine the temperature inside a ball of radius a if the boundary of the ball is maintained at temperature zero and the initial temperature is equal to $u|_{t=0} = \sin\left(\dfrac{\pi \rho}{a}\right)$, where ρ denotes the distance from the center of the ball to the current interior point.

29. Determine the temperature inside a ball of radius a if the boundary of the ball is maintained at temperature $u|_{\rho=a} = P_2(\cos\theta)$ and the initial temperature is equal to zero; here $P_2(x)$ is the Legendre polynomial of second degree.

30. A ball of radius a, initially heated to the temperature

$$
u|_{t=0} = \rho^2 P_{43}(\cos\theta) \cos(3\varphi),
$$

cools down while its boundary is maintained at temperature zero. Find the temperature inside the ball; here $P_{43}(x)$ is associate Legendre function:

$$
P_{43}(x) = (1 - x^2)^{3/2} \frac{d^3 P_4(x)}{dx^3}.
$$

31. Solve the Cauchy problem

$$\begin{cases} u_t = \Delta u, & -\infty < x, y, z < \infty, \quad t > 0, \\ u|_{t=0} = \cos(x + y + z), & -\infty < x, y, z < \infty. \end{cases}$$

32. Solve the Cauchy problem

$$\begin{cases} u_t = 4u_{xx} + t + e^t, & -\infty < x < \infty, \quad t > 0, \\ u|_{t=0} = 2, & -\infty < x < \infty. \end{cases}$$

33. Solve the Cauchy problem

$$\begin{cases} u_t = \dfrac{1}{2}\Delta u, & -\infty < x, y < \infty, \quad t > 0, \\ u|_{t=0} = \cos(xy), & -\infty < x, y < \infty. \end{cases}$$

34. Solve the Cauchy problem

$$\begin{cases} u_t = 2\Delta u + t \cos x, & -\infty < x, y, z < \infty, \quad t > 0, \\ u|_{t=0} = \cos y \cos z, & -\infty < x, y, z < \infty. \end{cases}$$

35. Solve the Cauchy problem

$$\begin{cases} u_t = \dfrac{1}{2}\Delta u + t \sin x \cos y, & -\infty < x, y < \infty, \quad t > 0, \\ u|_{t=0} = xy, & -\infty < x, y < \infty. \end{cases}$$

36. Solve the Cauchy problem

$$\begin{cases} u_t = \Delta^2 u + f(x, t), & x \in \mathbf{R}^n, \quad t > 0, \\ u|_{t=0} = g(x), & x \in \mathbf{R}^n, \end{cases}$$

where $g(x)$ is a biharmonic function and $f(x, t)$ is a biharmonic function of x for each fixed t.

37. Solve the Cauchy problem

$$\begin{cases} u_t = \Delta^2 u, & x \in \mathbf{R}^n, \quad t > 0, \\ u|_{t=0} = \displaystyle\prod_{i=1}^{n} \sin(k_i x_i), & x \in \mathbf{R}^n. \end{cases}$$

38. Solve the Cauchy problem

$$\begin{cases} u_{tt} = \Delta^2 u, & x \in \mathbf{R}^n, \quad t > 0, \\ u|_{t=0} = \sin x_1, & u_t|_{t=0} = \cos x_n. \end{cases}$$

39. Find a solution, having the form of a travelling wave propagating with constant velocity a, for the nonlinear equation

$$u_t = \frac{\partial u}{\partial x}\left(x^\alpha \frac{\partial u}{\partial x}\right), \qquad \alpha \geq 1.$$

40. Show that the equation $u_t = u_{xx}$ is equivalent to the integral identity

$$\oint_L u\, dx + u_x\, dt = 0;$$

here the function $u(x,t)$ is assumed to be sufficiently smooth and L is a closed piecewise-smooth contour.

3.6. Answers

1. $u = \dfrac{1}{2a\sqrt{\pi t}} \displaystyle\int_{-\infty}^{\infty} f(\zeta)e^{-\frac{(x-\zeta)^2}{4a^2 t}}\, d\zeta.$

2. $u = \dfrac{1}{2a\sqrt{\pi t}} \displaystyle\int_{-\infty}^{\infty} f(\zeta)\left[e^{-\frac{(x-\zeta)^2}{4a^2 t}} - e^{-\frac{(x+\zeta)^2}{4a^2 t}}\right] d\zeta.$

3. $u = \dfrac{1}{(2a\sqrt{\pi t})^2} \displaystyle\int_{-\infty}^{\infty} d\xi \int_{-\infty}^{\infty} d\eta \int_{-\infty}^{\infty} f(\xi,\eta,\zeta)\times$

$$\times \left[e^{-\frac{(x-\xi)^2+(y-\eta)^2+(z-\zeta)^2}{4a^2 t}} - e^{-\frac{(x-\xi)^2+(y-\eta)^2+(z+\zeta)^2}{4a^2 t}}\right] d\zeta.$$

5. $u = -\dfrac{1}{a}\displaystyle\int_0^x f(x-\xi)\sin(a\xi)d\xi.$

6. $u = be^{-3t}\cos(2x) - \dfrac{a}{2}x\sin x.$

7. $u = \dfrac{x}{2a\sqrt{\pi}\, t^{3/2}}e^{-\frac{x^2}{4a^2 t}}.$

8. $u = 2e^{-9a^2 t}\sin(3x) + 5e^{-64a^2 t}\sin(8x).$

9. $u = 2e^{-a^2 t}\cos x + 3e^{-4a^2 t}\cos(2x).$

10. $u = \dfrac{1}{4a^2}\left[t + \dfrac{1}{4a^2}\left(e^{-4a^2 t} - 1\right)\right].$

12. $u = 3e^{-26a^2 t}\sin x \sin(5y).$

13. $u = 5e^{-10a^2 t}\cos x \cos(3y).$

14. $u = \dfrac{A}{1+4a^4}\left(e^{-2a^2 t} - 2a^2 t + 2a^2\sin t - \cos t\right)\sin x \sin y.$

15. $u = \dfrac{\sinh x}{\sinh \pi}\sin y + \dfrac{2}{\pi}\displaystyle\sum_{n=1}^{\infty}\dfrac{n(-1)^n}{n^2+1}e^{-a^2(n^2+1)t}\sin(nx)\sin y.$

18. $u = At\dfrac{l-x}{l} - \dfrac{2Al^2}{\pi^3 a^2}\displaystyle\sum_{n=1}^{\infty}\dfrac{1}{n^3}\left(1 - e^{-a^2\left(\frac{n\pi}{l}\right)^2 t}\right)\sin\left(\dfrac{n\pi}{l}x\right).$

19. $u = Ae^{-a^2\left(\frac{\mu_k}{R}\right)^2 t}J_0\left(\dfrac{\mu_k}{R}\rho\right).$

20. $u = u_0\left[1 + 2\displaystyle\sum_{n=1}^{\infty}e^{-a^2\left(\frac{\mu_n}{R}\right)^2 t}\dfrac{J_0\left(\frac{\mu_n}{R}\rho\right)}{\mu_n J_0'(\mu_n)}\right].$

21. $u = Ae^{-a^2\left(\frac{\mu_3}{a}\right)^2 t}J_2\left(\dfrac{\mu_3}{a}\rho\right)\cos(2\varphi),$

where c is the constant appearing in the equation $u_t = c^2\Delta u.$

22. $u = A\dfrac{1}{c^2}\left(\dfrac{a}{\mu_1}\right)^2\left[1 - e^{-c^2\left(\frac{\mu_1}{a}\right)^2 t}\right]J_0\left(\dfrac{\mu_1}{a}\rho\right).$

23. $u = \dfrac{1}{1 + c^2\sqrt{\left(\frac{\mu_1}{a}\right)^2 + \left(\frac{3\pi}{h}\right)^2}}\left[e^t - e^{-c^2\sqrt{\left(\frac{\mu_1}{a}\right)^2 + \left(\frac{3\pi}{h}\right)^2}\,t}\right]\times$

$$\times J_0\left(\dfrac{\mu_1}{a}\rho\right)\cos\left(\dfrac{3\pi}{h}z\right).$$

24. $u = -A\cos(2\varphi)\displaystyle\sum_{m=1}^{\infty}A_m J_2\left(\dfrac{\mu_m}{a}\rho\right)e^{-c^2\left(\frac{\mu_m}{a}\right)^2 t} + A\dfrac{\rho^2}{a^2}\cos(2\varphi),$

where μ_m is the mth positive root of the Bessel function $J_2(x)$ and the coefficients of the series are given by the formula

$$A_m = \dfrac{2A\displaystyle\int_0^a \rho^3 J_2\left(\frac{\mu_m}{a}\rho\right)d\rho}{a^4[J_2'(\mu_m)]^2}.$$

25. $u = A\dfrac{J_0\left(\frac{\omega}{c}\rho\right)}{J_0\left(\frac{\omega}{c}a\right)}e^{-\omega^2 t} - \displaystyle\sum_{n=1}^{\infty}A_n J_0\left(\dfrac{\mu_n}{a}\rho\right)e^{-c^2\left(\frac{\mu_n}{a}\right)^2 t},$ where

$$A_n = \dfrac{2A\displaystyle\int_0^a \rho J_0\left(\frac{\omega}{c}\rho\right)J_0\left(\frac{\mu_n}{a}\rho\right)d\rho}{a^2 J_0\left(\frac{\omega}{c}\rho\right)[J_1(\mu_n)]^2},$$

c is the constant appearing in the equation $u_t = c^2\Delta u$, and μ_n is the nth positive root of the equation $J_0(x) = 0.$

26. $u = A \dfrac{I_0\left(\frac{\omega}{c}\rho\right)}{I_0\left(\frac{\omega}{c}a\right)} e^{\omega^2 t} - \sum_{n=1}^{\infty} A_n J_0\left(\dfrac{\mu_n}{a}\rho\right) e^{-c^2\left(\frac{\mu_n}{a}\right)^2 t}$, where

$$A_n = \dfrac{2A \displaystyle\int_0^a \rho I_0\left(\frac{\omega}{c}\rho\right) I_0\left(\frac{\mu_n}{a}\rho\right) d\rho}{a^2 I_0\left(\frac{\omega}{c}\rho\right) [J_1(\mu_n)]^2}$$

and $I_0(x)$ is the Bessel function of index zero and imaginary argument.

References

[1] Arsenin, V. Ya., *Methods of Mathematical Physics, and Special Functions,* "Nauka," Moscow, 1974. (Russian)

[2] Bers, L., John, F., and Schechter, M., *Partial Differential Equations,* Reprint of the 1964 original. Lectures in Applied Mathematics, 3A. American Mathematical Society, Providence, R.I., 1979.

[3] Bitsadze, A. V. and Kalinichenko, D. F., *Collection of Problems on the Equations of Mathematical Physics,* Second edition, "Nauka", Moscow, 1985. (Russian)

[4] Budak, B. M., Samarskiĭ, A. A., and Tikhonov, A. N., *A Collection of Problems in Mathematical Physics,* Third edition, "Nauka", Moscow, 1980. (Russian)

[5] Vladimirov, V. S., *Equations of Mathematical Physics,* Fifth edition, "Nauka", Moscow, 1988. (Russian); English transl., *Equations of Mathematical Phsyics,* "Mir", Moscow, 1984.

[6] Vladimirov, V. S., Mikhailov, V. P., Vasharin, A. A., Karimova, Kh. Kh., Sidorov, Yu. V., and Shabunin, M. I., *A Collection of Problems on the Equations of Mathematical Physics,* "Nauka", Moscow, 1974. (Russian); English transl., "Mir", Moscow; Springer- Verlag, Berlin-New York, 1986.

[7] Courant, R., *Partial Differential Equations,* Translated from the English by T. D. Ventcel'; edited by O. A. Oleĭnik "Mir", Moscow, 1964. (Russian); Translation of: Courant, R. and Hilbert, D., *Methods of Mathematical Physics. Vol. II: Partial Differential Equations,* Interscience Publishers (a division of John Wiley & Sons), New York–London, 1962.

[8] Courant, R. and Hilbert, D., *Methods of Mathematical Physics. Vol. II. Partial differential equations,* Reprint of the 1962 original. Wiley Classics Library. A Wiley-Interscience Publication. John Wiley & Sons, Inc., New York, 1989.

[9] Lavrent'ev, M. A. and Shabat, B. V., *Methods of the Theory of Functions of a Complex Variable,* Fifth edition. "Nauka", Moscow, 1987. (Russian)

[10] Ladyzhenskaya, O. A., *Boundary ·Value Problems of Mathematical Physics,* "Nauka", Moscow, 1973. (Russian); English transl., *The Boundary Value Problems of Mathematical Physics,* Translated by Jack Lohwater, Applied Mathematical Sciences, 49. Springer- Verlag, New York-Berlin, 1985.

[11] Maslov, V. P., *Asymptotic Methods and Perturbation Theory,* Nauka", Moscow, 1988. (Russian)

[12] Mikhlin, S. G. *Linear Partial Differential Equations,* Izdat. "Vyshch. Shkola", Moscow, 1977. (Russian); German transl., *Partielle Differentialgleichungen in der Mathematischen Physik,* Translated by Bernd Silbermann, Mathematische Lehrbücher und Monographien, I. Abteilung: Mathematische Lehrbücher, 30. Akademie-Verlag, Berlin, 1978.

[13] Nayfeh, Ali Hasan, *Perturbation Methods,* Pure and Applied Mathematics. John Wiley & Sons, New York-London-Sydney, 1973.

[14] Nikiforov, A. F. and Uvarov, V. B., *Special Functions of Mathematical Physics,* "Nauka", Moscow, 1978. (Russian); Englis transl., *Special Functions of Mathematical Physics. A Unified Introduction with Applications,* Translated and with a preface by Ralph P. Boas, Birkhäuser Verlag, Basel-Boston, MA, 1988.

[15] Pikulin, V. P., *Elaboration of Methods for Equations of Elliptic and Parabolic Type,* Moscow Power Institute, Moscow, 1990. (Russian)

[16] Pikulin, V. P. and Pohozaev, S. I., *Practical Course on Equations of Mathematical Physics,* Moscow Power Institute, Moscow, 1989. (Russian)

[17] Sidorov, Yu. V., Fedoryuk, M. V. and Shabunin, M. I., *Lectures on the Theory of Functions of a Complex Variable,* Third edition. "Nauka", Moscow, 1989. (Russian); English transl., *Lectures on the Theory of Functions of a Complex Variable,* Translated by Eugene Yankovsky, "Mir", Moscow, 1985.

[18] Smirnov, M. M., *Problems on the Equations of Mathematical Physics,* Sixth edition, supplemented, "Nauka", Moscow, 1975. (Russian); English transl., *Problems on the Equations of Mathematical Physics,* Translated by W. I. M. Wils, Noordhoff, Groningen, 1966.

[19] Sobolev, S. L. *Equations of Mathematical Physics,* Fourth edition, "Nauka", Moscow, 1966. (Russian); English transl., *Partial Differential Equations of Mathematical Physics,* Translated from the third Russian edition by E. R. Dawson, Pergamon Press, Oxford-Edinburgh-New York-Paris-Frankfurt; Addison-Wesley Publishing Co., Inc., Reading, Mass.-London, 1964.

[20] Tikhonov, A. N., Vasil'eva, A. B. and Sveshnikov, A. G. *Differential Equations,* Course in Higher Mathematics and Mathematical Physics, 7, "Nauka", Moscow, 1980. (Russian); English transl., *Differential Equations,* Translated by A. B. Sosinskiĭ, Springer Series in Soviet Mathematics. Springer-Verlag, Berlin-New York, 1985.

[21] Tikhonov, A. N. and Samarskiĭ, A. A., *Equations of Mathematical Physics,* Fourth edition, corrected. Izdat. "Nauka", Moscow, 1972. (Russian); English transl., *Equations of Mathematical Physics,* Translated by A. R. M. Robson and P. Basu, A Pergamon Press Book. The Macmillan Co., New York 1963; Reprint of the 1963 translation. Dover Publications, Inc., New York, 1990.

[22] Farlow, Stanley J., *Partial Differential Equations for Scientists and Engineers,* John Wiley & Sons, Inc., New York, 1982; Revised reprint, Dover Publications, Inc., New York, 1993.

[23] Fedoryuk, M. V., *Asymptotic Methods for Linear Ordinary Differential Equations,* Mathematical Reference Library, "Nauka", Moscow, 1983. (Russian)

[24] Fedoryuk, M. V., *Ordinary Differential Equations,* Selected Chapters in Higher Mathematics for Engineers and Students in Higher Technical Scho-ols, "Nauka", Moscow, 1980. (Russian)

Index